Study Guide

T0202253

Chemistry Unit 1 for CAPE®

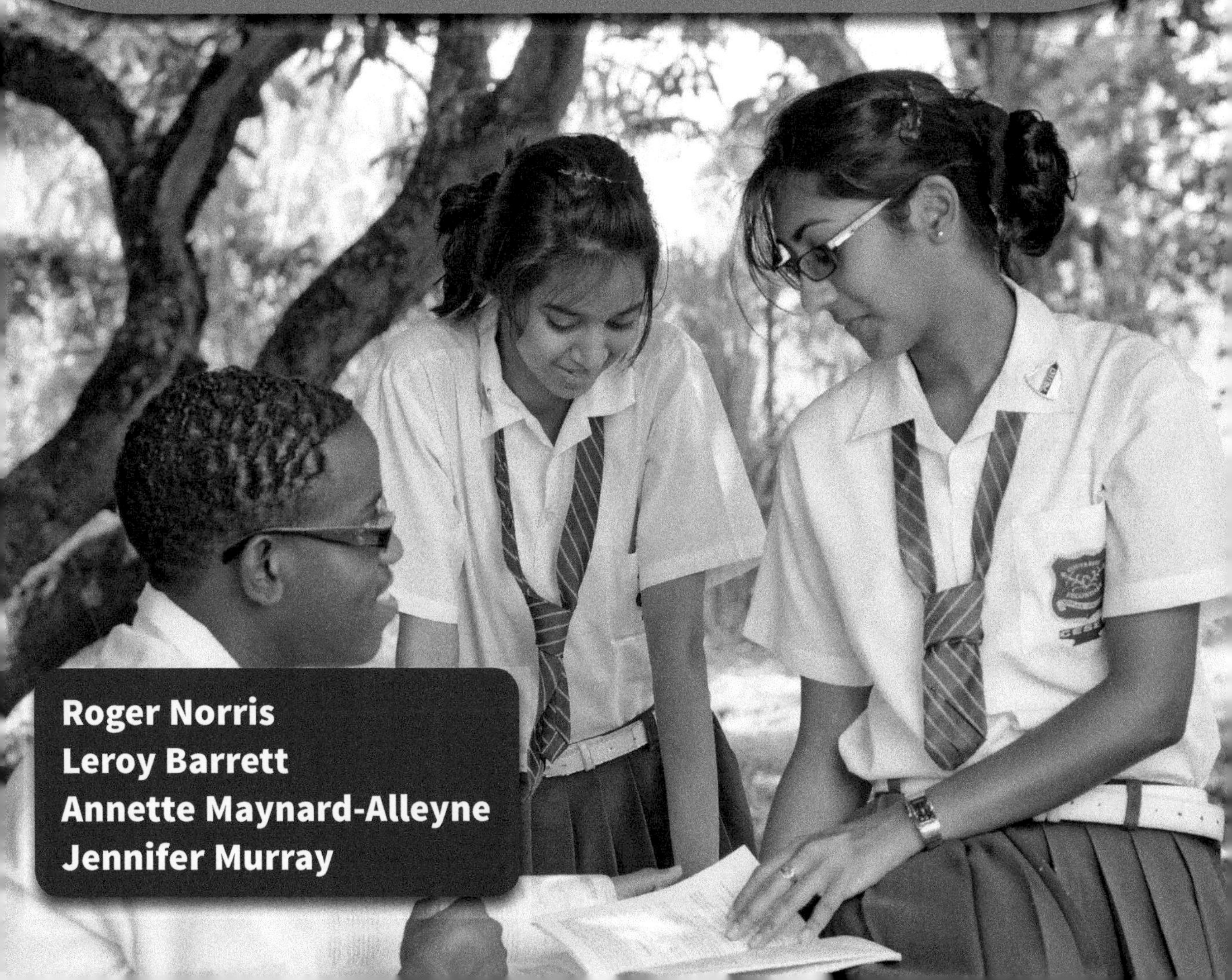

Roger Norris
Leroy Barrett
Annette Maynard-Alleyne
Jennifer Murray

OXFORD
UNIVERSITY PRESS

Great Clarendon Street, Oxford, OX2 6DP, United Kingdom

Oxford University Press is a department of the University of Oxford. It furthers the University's objective of excellence in research, scholarship, and education by publishing worldwide. Oxford is a registered trade mark of Oxford University Press in the UK and in certain other countries

First published by Nelson Thornes Ltd in 2012
This edition published by Oxford University Press in 2015

British Library Cataloguing in Publication Data
Data available

978-1-4085-1668-3

14

Printed in Great Britain by CPI Group (UK) Ltd., Croydon CR0 4YY

Acknowledgements

Cover: Mark Lyndersay, Lyndersay Digital, Trinidad
www.lyndersaydigital.com
Illustrations: Wearset Ltd, Boldon, Tyne and Wear
Page make-up: Wearset Ltd
Index: Indexing Specialists (UK) Ltd

Photographs
Page 2 iStockphoto

Contents

Introduction

This Study Guide has been developed exclusively with the Caribbean Examinations Council (CXC®) to be used as an additional resource by candidates, both in and out of school, following the Caribbean Advanced Proficiency Examination (CAPE®) programme.

It has been prepared by a team with expertise in the CAPE® syllabus, teaching and examination. The contents are designed to support learning by providing tools to help you achieve your best in CAPE® Chemistry and the features included make it easier for you to master the key concepts and requirements of the syllabus. *Do remember to refer to your syllabus for full guidance on the course requirements and examination format!*

Inside this Study Guide is an interactive CD, which includes electronic activities to assist you in developing good examination techniques:

- **On Your Marks** activities provide sample examination-style short answer and essay type questions, with example candidate answers and feedback from an examiner to show where answers could be improved. These activities will build your understanding, skill level and confidence in answering examination questions.
- **Test Yourself** activities are specifically designed to provide experience of multiple-choice examination questions and helpful feedback will refer you to sections inside the study guide so that you can revise problem areas.

This unique combination of focused syllabus content and interactive examination practice will provide you with invaluable support to help you reach your full potential in CAPE® Chemistry.

1 Atomic structure

1.1 The structure of the atom

Learning outcomes

On completion of this section, you should be able to:

- describe Dalton's atomic theory
- describe the structure of the atom in terms of position, relative charges and masses of the sub-atomic particles.

John Dalton (1766–1844)

Did you know?

Scientists now think that matter is made up of sub-atomic particles called 'leptons' and 'quarks'. An electron is a type of lepton and protons and neutrons are made up of quarks.

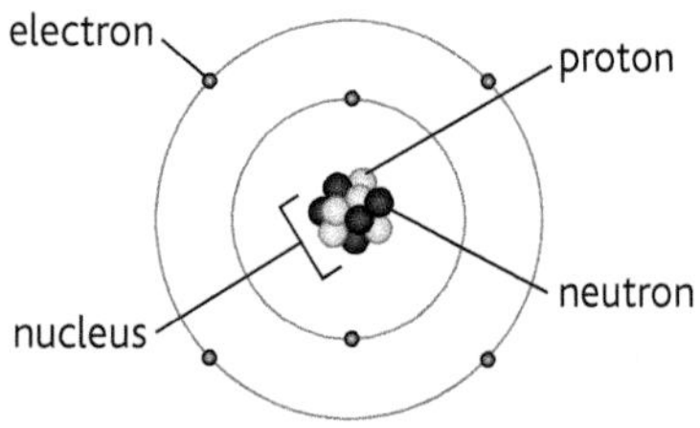

Figure 1.1.2 *The nucleus of an atom contains protons and neutrons. The electrons are outside the nucleus in fixed energy levels.*

Dalton's atomic theory

John Dalton thought of atoms as being hard spheres. He suggested that:

- all atoms of the same element are exactly alike
- atoms cannot be broken down any further
- atoms of different elements have different masses
- atoms combine to form more complex structures (**compounds**).

Later changes to Dalton's atomic theory

Dalton's atomic theory was first described in 1807. Over the past two centuries, scientists have modified the theory to fit the evidence available.

1897

Thomson discovered the electron (which he called a 'corpuscle'). He suggested the 'plum-pudding' model of the atom: electrons embedded in a sea of positive charge.

1909

Rutherford suggested that most of the mass of the atom was in a tiny positively-charged nucleus in the middle of the atom. This led to the planetary model with electrons surrounding the nucleus.

1913

Bohr suggested that electrons could only orbit the nucleus at certain distances depending on their energy.

1932

Chadwick discovered the presence of the neutron in the nucleus.

1926 onwards

Schrödinger, Born and Heisenberg described the position of an electron in terms of the probability of finding it at any one point. This led to the idea of atomic orbitals (Section 1.4).

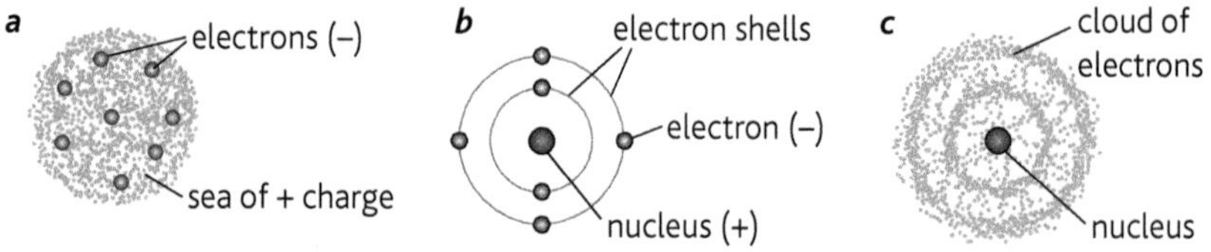

Figure 1.1.1 *Three different models of the atom:* ***a*** *The plum-pudding model;* ***b*** *The Bohr model;* ***c*** *The atomic orbital model*

Figure 1.1.2 shows a useful model of an atom based on the Bohr model.

Masses and charges of sub-atomic particles

The actual masses and charges of protons, neutrons and electrons are very small. For example, a proton has a mass of 1.67×10^{-27} kg and a charge of $+1.6 \times 10^{-19}$ C. The relative masses and charges give an easier comparison. These are shown in the table.

	Proton	**Neutron**	**Electron**
Relative mass	1	1	$\frac{1}{1836}$
Relative charge	+1	0	–1

Sub-atomic particles and electric fields

The effect of a strong electric field on beams of protons, neutrons and electrons is shown below.

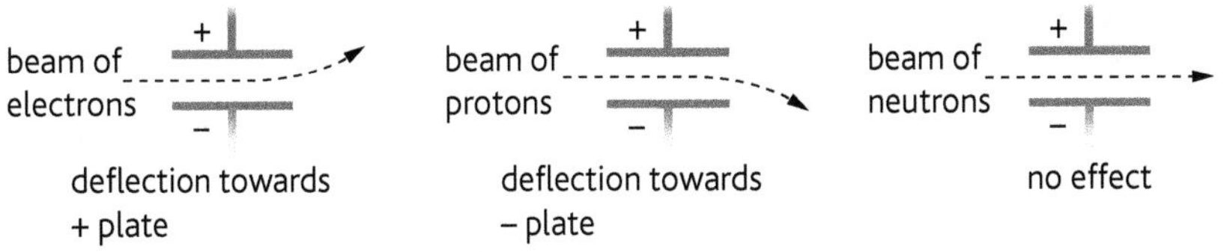

Figure 1.1.3 *The effect of an electric field on beams of electrons, protons and neutrons*

Charged particles move towards the plates with opposite charge and away from the plates with the same charge. If we use the same voltage to deflect electrons and protons, the electrons are deflected to a much greater extent. This is because electrons have a very small mass compared with protons.

Sub-atomic particles and magnetic fields

Figure 1.1.4 shows an evacuated glass tube. When heated by a low voltage supply, electrons are produced from the tungsten filament, F. They are accelerated towards the metal cylinder, C. They produce a glow when they collide with the fluorescent screen. When the north pole of a magnet is brought near the electron beam, the glow on the fluorescent screen moves downwards. The direction of the movement can be predicted from Fleming's left-hand rule. The direction of movement is at right angles to the direction of the magnetic field and the conventional current. Remember that the direction of electron flow is opposite that of the conventional current.

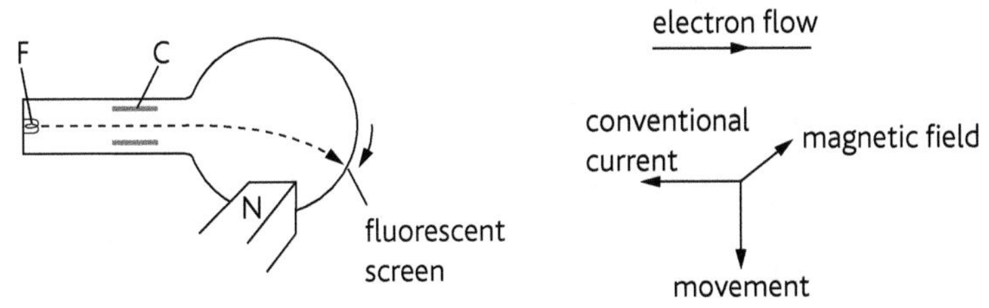

Figure 1.1.4 *The effect of a magnetic field on a beam of electrons*

Using different apparatus, beams of protons will be deflected in the opposite direction to the electrons. Beams of neutrons will not be deflected since they have no charge.

Exam tips

The direction of the movement of an electron in a magnetic field can be found by Fleming's left-hand rule as shown below. Remember (1) The direction of the magnetic field is N → S (2) that the centre finger points in the direction **opposite** to the electron flow.

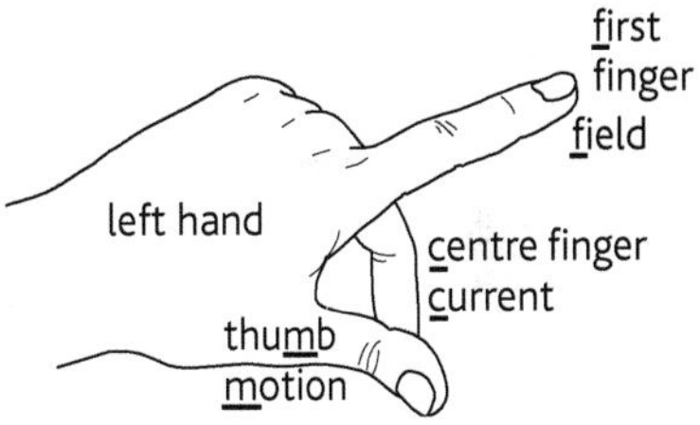

Key points

- The atomic theory has developed to fit the experimental evidence.
- Most of the mass of the atom is in the nucleus which contains protons and neutrons. The electrons are outside the nucleus in shells.
- Electrons, protons and neutrons have characteristic relative masses and charges.
- Beams of protons and electrons are deflected in opposite directions by electric and magnetic fields. Neutrons are not deflected.

1.2 Isotopes and radioactivity

Learning outcomes

On completion of this section, you should be able to:

- define 'mass number' and 'isotopes'
- define 'relative atomic mass' and 'relative isotopic mass' using the ^{12}C scale
- know how to calculate relative atomic masses
- write equations involving α-, β- and γ-radiation
- state some uses of radioisotopes.

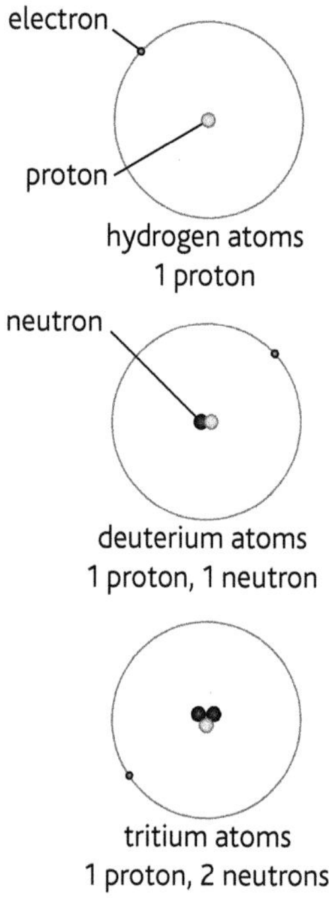

***Figure 1.2.1** The three isotopes of hydrogen*

Did you know?

Isotopes with 2, 8, 20, 28, 50, 82 and 126 neutrons or protons are extremely stable. These 'magic numbers' are thought to indicate closed nuclear shells comparable to the full shells of the noble gases.

Atomic number and mass number

Atomic number (*Z*)

Atomic number (*Z*) is the number of protons in the nucleus of an atom. (It also equals the number of electrons in a neutral atom and the position of the element in the Periodic Table.)

Mass number (*A*)

Mass number (*A*) is the number of protons plus neutrons in the nucleus of an atom.

So the number of neutrons in an atom is $A - Z$ (mass number − atomic number).

Isotopes

Isotopes are atoms with the same atomic number but different mass numbers.

We show the mass number and atomic number of an isotope like this:

$$\begin{matrix}\text{mass number} \rightarrow \\ \text{atomic number} \rightarrow\end{matrix} {}^{23}_{11}\text{Na} \leftarrow \text{element symbol}$$

Relative atomic and relative isotopic mass

The mass of a single atom is too small to weigh, so we use the ^{12}C (carbon-12) atom as a standard. We compare the masses of other atoms to this standard, which has a mass of exactly 12 units.

Relative atomic mass (A_r)

Relative atomic mass (A_r) is the weighted average mass of naturally occurring atoms of an element on a scale where an atom of ^{12}C has a mass of exactly 12 units.

Relative isotopic mass

Relative isotopic mass is the mass of a particular isotope of an element on a scale where an atom of ^{12}C has a mass of exactly 12 units.

Calculating accurate relative atomic masses

Most naturally occurring elements exist as a mixture of isotopes. So their mass numbers are not always exact. We have to take into account the proportion of each isotope. This is called its **relative isotopic abundance**. The relative atomic mass for neon is calculated below from the relative isotopic masses and relative abundance.

Isotope	Relative isotopic mass	Relative abundance/%
^{20}Ne	20	90.9
^{21}Ne	21	0.30
^{22}Ne	22	8.8

$$A_r = \frac{(20.0 \times 90.9) + (21.0 \times 0.3) + (22 \times 8.8)}{100} = 20.2$$

Radioactivity

Isotopes of some elements have nuclei which break down (decay) spontaneously. These are described as **radioactive isotopes**. As the nuclei break down, rays or particles are given out. These are called emissions. The table shows three types of emission.

Name of emission	Type of particles/rays emitted	Stopped by
Alpha (α)	helium nuclei (positively charged particles)	thin sheet of paper
Beta (β)	electrons (produced by **nuclear** changes)	6 mm thick aluminium foil
Gamma (γ)	very high frequency electromagnetic radiation	thick lead sheet

Equations can be written for each of these types of decay. For example:

α-decay
The isotope produced has a mass number of 4 units lower and a nuclear charge of 2 units lower than the original atom:

$$^{223}_{88}Ra \rightarrow {}^{219}_{86}Rn + {}^{4}_{2}He$$

β-decay
The mass number stays the same but the number of protons increases by one. This is because a neutron (^1_0n) is changed into a proton (^1_1p) and an electron ($^{\ 0}_{-1}e$):

$$({}^1_0n) \rightarrow ({}^1_1p) + ({}^{\ 0}_{-1}e)$$

For example carbon-14 decays to nitrogen-14 when it emits a β-particle:

$$^{14}_{6}C \rightarrow {}^{14}_{7}N + {}^{\ 0}_{-1}e$$

γ-decay
γ-rays can be emitted along with α- or β-particles or in a process called 'electron capture'. A proton is converted to a neutron, so the mass number stays the same but the atomic number decreases by one. For example:

$$^{37}_{18}Ar + {}^{\ 0}_{-1}e \rightarrow {}^{37}_{17}Cl$$

Uses of radioisotopes

- Tracers for searching for faults in pipelines and for studying the working of certain organs in the body, e.g. ^{131}I is used to study thyroid function.
- In medicine, for radiotherapy in the treatment of cancers.
- Dating the age of objects, e.g. using ^{14}C to date objects which were once living.
- Smoke detectors often use ^{241}Am.
- Generating power, e.g. ^{235}U is used in many nuclear reactors as a source of energy.

Exam tips

A common error is to suggest that α-particles are helium atoms. They are not: they are helium nuclei He^{2+}. Another error is to think that β-particles are electrons from the outside the nucleus. The β-particle electrons arise from the nucleus not from the electron shells.

Key points

- Mass number is the total number of protons + neutrons in an isotope.
- Isotopes are atoms with the same number of protons but different numbers of neutrons.
- Relative atomic and isotopic masses are compared using the ^{12}C scale.
- The relative atomic mass of an element is the weighted mean of the atoms of the isotopes they contain.
- When an unstable isotope decays rays or particles are emitted.
- α-, β-, and γ-radiation have characteristic properties.

1.3 Energy levels and emission spectra

Learning outcomes

On completion of this section, you should be able to:

- explain how data from emission spectra give evidence of discrete energy levels for electrons
- explain the lines of the Lyman and Balmer series in emission spectra
- understand how the difference in energy between two electrons' energy levels is related to the frequency of radiation.

Energy quanta and energy levels

The electrons in atoms have certain fixed values of energy. The smallest fixed amount of energy required for a change is called a **quantum** of energy. We say that the energy is discrete or **quantised**. The electrons are arranged in energy levels which are quantised. The atom is most stable when the electrons are in the lowest energy levels possible for each of them. The electrons are then in their **ground state**. When an electron absorbs a quantum of radiation it can move up to a higher energy level. The electron is then said to be in an **excited state**. When an excited electron falls to a lower energy level a quantum of radiation is given out. The electron moves to a lower energy level.

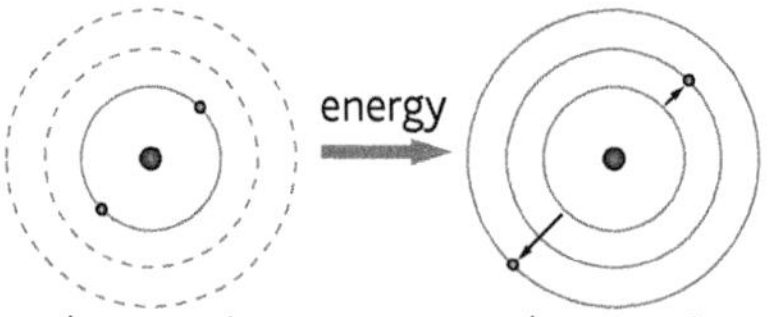

Figure 1.3.1 *An electron can jump from one energy level to another by absorbing or emitting a fixed amount (quantum) of energy*

The difference in energy between two energy levels is related to the frequency of radiation by the relationship:

$$\text{energy (J)} \rightarrow \Delta E = h\nu \leftarrow \text{frequency of radiation (s}^{-1}\text{)}$$

$$\uparrow$$

$$\text{Planck's constant } (6.63 \times 10^{-34}\,\text{J s}^{-1})$$

In any atom there are several possible energy levels. These are called **principal quantum levels** or **principal energy levels**. They are also called electron shells. They are given numbers $n = 1$, $n = 2$, $n = 3$, etc. going further away from the nucleus.

Atomic emission spectra

When electrical or thermal energy is passed through a gaseous sample of an element, the radiation is emitted only at certain wavelengths or frequencies. An **emission spectrum** differs from a normal visible light spectrum in that:

- It is made up of separate lines.
- The lines **converge** (get closer to each other) as their frequency increases.

Figure 1.3.2 *Part of a simplified line emission spectrum of hydrogen*

Interpreting the hydrogen emission spectrum

Each line in the hydrogen emission spectrum is a result of electrons moving from a higher to a lower energy level. Among the several series of lines seen are the:

- **Lyman series** (seen in the ultraviolet region), where previously excited electrons fall back to the $n = 1$ energy level.
- **Balmer series** (seen in the visible region), where previously excited electrons fall back to the $n = 2$ energy level.

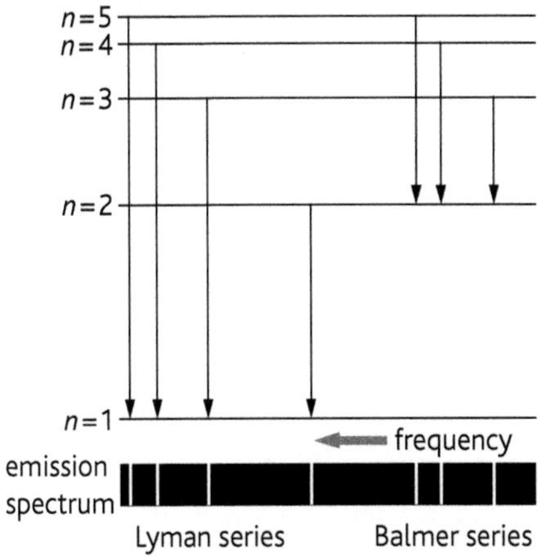

Figure 1.3.3 *A line emission spectrum is the result of electrons falling from higher energy levels to lower energy levels*

The energy associated with each line is found by using the relationship $\Delta E = h\nu$. You can see that the greater the frequency (the smaller the wavelength), the more energy is released. The point where the lines eventually come together is called the **convergence limit**. This represents an electron falling from the highest possible energy level. If the electron has more energy than this, it becomes free from the pull of the nucleus of the atom. Then the atom is converted to an ion.

Did you know?

There are other sets of emission spectrum lines in the infrared region. These are called the Paschen, Brackett and Pfund series.

The Bohr model of an atom

The energy levels get closer together towards the outside of the atom. This is mirrored by the spectral lines getting closer together. In the Bohr model of the hydrogen atom shown opposite, a quantum of energy moves the hydrogen electron in the $n = 1$ level (ground state) to the $n = 2$ level. A further quantum of energy will excite the electron to the $n = 3$ level. The more energy an electron has, the higher the energy level it will be in. When the electron loses energy, it will fall down to the lower energy levels, emitting radiation of characteristic frequencies.

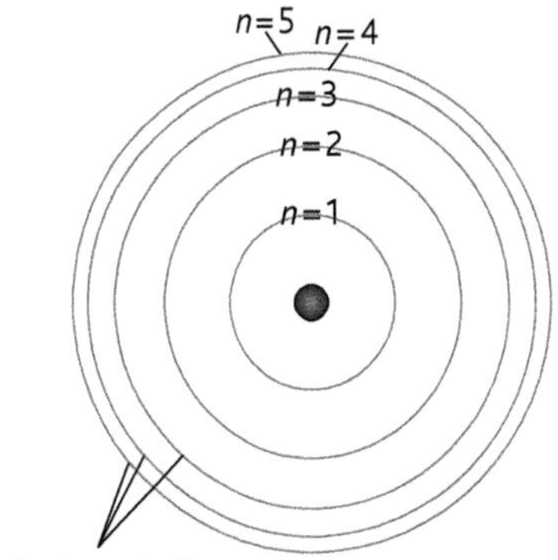

Figure 1.3.4 *The Bohr model of the atom. The principal energy levels get closer together towards the outside of the atom.*

Key points

- Electrons occupy specific energy levels (quantum levels) in the atom.
- When electrons gain specific quanta of energy they move from lower to higher energy levels. They become 'excited'.
- When excited, electrons lose energy, they fall back to lower energy levels emitting radiation of characteristic frequency. This is the origin of the line emission spectrum.
- The energy difference between any two energy levels is given by $\Delta E = h\nu$.

Sub-shells and atomic orbitals

Learning outcomes

On completion of this section, you should be able to:

- know that the principal quantum shell for $n = 2$ and above can be split into sub-shells (subsidiary quantum shells)
- know the relative energies of the shells and sub-shells
- know how to determine the electronic configuration of atoms and ions
- understand the term 'atomic orbital'
- describe the shapes of s and p orbitals.

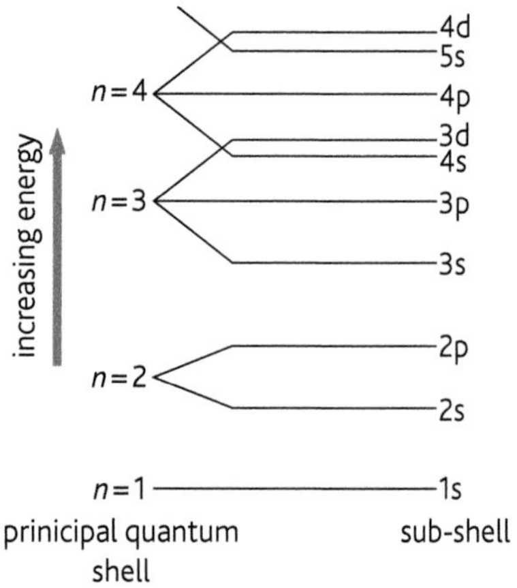

Figure 1.4.1 *The principal quantum shells, apart from n = 1, are split into sub-shells*

Shells and sub-shells

The principal quantum shells apart from the first, are split into **sub-shells** (sub-levels). These are named s, p, d. Figure 1.4.1 shows the sub-shells for the first four quantum shells in terms of increasing energy. Notice that after the element calcium the order of the sub-shells, in terms of energy, does not always conform to the pattern of s followed by p followed by d. For example in scandium the 4s sub-shell is at a lower energy level than the 3d.

For a neutral atom, the number of electrons is equal to the number of protons. The *maximum* number of electrons each shell and sub-shell can hold is shown in the table.

Principal quantum shell	Total number of electrons in principal quantum shell	Maximum number of electrons in sub-shell		
		s	p	d
$n = 1$	2	2	–	–
$n = 2$	8	2	6	–
$n = 3$	18	2	6	10

Atomic orbitals

Each sub-shell contains one or more **atomic orbitals**. An orbital is a region of space where there is a high probability of finding an electron. Each orbital can hold a *maximum* of two electrons. So the number of orbitals in each sub-shell is: s = 1, p = 3 and d = 5. Orbitals differ from each other in shape (see Figure 1.4.2).

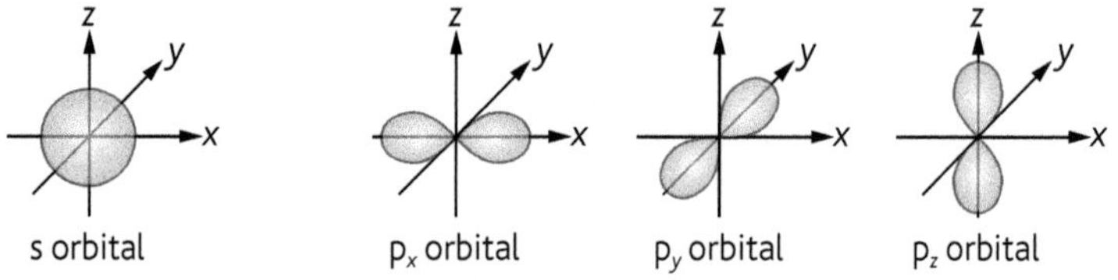

Figure 1.4.2 *The shapes of s and p orbitals*

Note that:

- A 2s orbital is larger than the 1s orbital and a 3p is larger than the 2p.
- The three different p orbitals are mutually at right angles to each other.

Here are two examples of how to write the type of orbital and the number of electrons present in them:

2 electrons in orbital ↓
First quantum shell → $1s^2$
s sub-shell ↑

5 electrons ↓
Third quantum shell → $3p^5$
p sub-shell ↑

The electronic configuration of atoms

The **electronic configuration** shows the number and type of the electrons in a particular atom. The procedure is:

1 Make sure you know the order of the orbitals, 1s 2s 2p 3s etc., as well as the number of electrons in the atom.

2 Fill up the sub-shells to the maximum amount starting with those of lowest energy, e.g. an s sub-shell can have a maximum of 2 electrons and a p sub-shell a maximum of 6.

3 Stop when you have the correct number of electrons.

Example 1
Electronic configuration of sodium ($Z = 11$):

$$1s^2\ 2s^2\ 2p^6\ 3s^1$$

Example 2
Electronic configuration of scandium ($Z = 21$):

$$1s^2\ 2s^2\ 2p^6\ 3s^2\ 3p^6\ 4s^2\ 3d^1 \text{ (or } 1s^2\ 2s^2\ 2p^6\ 3s^2\ 3p^6\ 3d^1\ 4s^2\text{)}$$

Note that the 4s sub-shell is filled before the 3d (see Figure 1.4.1).

The electronic configuration of ions

When ions are formed one or more electrons are removed from or added to the atom. In general, the electrons are lost or gained from the outer sub-shell.

Example 1
Magnesium atom $1s^2\ 2s^2\ 2p^6\ 3s^2$; Magnesium ion (Mg^{2+}) $1s^2\ 2s^2\ 2p^6$

Example 2
Nitrogen atom $1s^2\ 2s^2\ 2p^3$; Nitride ion (N^{3-}) $1s^2\ 2s^2\ 2p^6$

Note that the electronic configuration of some transition element atoms and ions are less straightforward (see Section 13.1).

Exam tips

This diagram may help you remember the order of filling the sub-shells.

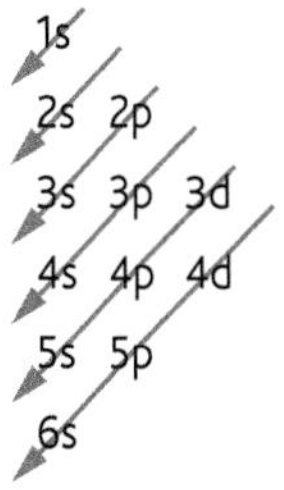

Did you know?

The electronic configuration can also be written to show the separate orbitals p_x, p_y, p_z. (see Figure 1.4.2). For example, the electronic configuration of sodium (Example 1) can be written:

$$1s^2\ 2s^2\ 2p_x^2\ 2p_y^2 p_z^2\ 3s^1$$

Key points

- The principal quantum shells (apart from the first) are divided into sub-shells named s, p, d, etc.
- Each sub-shell can hold a maximum number of electrons.
- An orbital is a region of space where there is a good chance of finding an electron.
- s orbitals are spherical in shape and p have an hour-glass shape.
- Each orbital can hold a maximum of 2 electrons.
- The maximum number of electrons in each sub-shell is s = 2, p = 6, d = 10.
- Electronic configurations of atoms and ions are found by adding electrons to the energy sub-levels placed in order of increasing energy.

Learning outcomes

On completion of this section, you should be able to:

- know the factors influencing ionisation energy
- explain how ionisation energy data give evidence for sub-shells
- know how electron configuration can be determined from successive ionisation energies.

Ionisation energy

The first **ionisation energy**, ΔH_{i1}, is the energy needed to remove one electron from each atom in one mole of atoms of an element in its gaseous state to form one mole of gaseous ions. For example:

$$Cl(g) \rightarrow Cl^{+}(g) + e^{-} \quad \Delta H_{i1} = +1260\,kJ\,mol^{-1}$$

The value of the ionisation energy depends on:

- **Size of nuclear charge:** In any one period, the higher the nuclear charge (greater number of protons), the greater the attractive force between the nucleus and outer electrons and the more energy needed to remove these electrons. So ΔH_{i1} tends to increase with increase in nuclear charge.
- **Distance of the outer electrons from the nucleus:** The further the outer electrons are from the nucleus, the less attractive force there is between them and the nucleus and the lower the value of ΔH_{i1}.
- **Shielding:** Electrons in full inner shells reduce the attractive force between the nucleus and the outer electrons. The greater the number of full inner shells, the greater the shielding and the lower the value of ΔH_{i1}.

First ionisation energies across a period

Figure 1.5.1 shows how the values of ΔH_{i1} change across the first three periods.

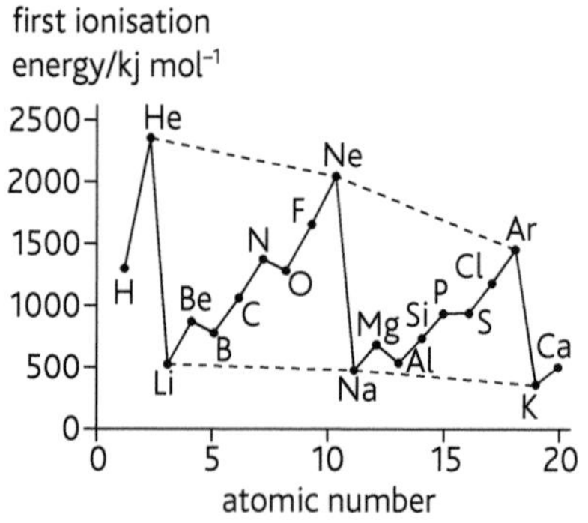

Figure 1.5.1 *First ionisation energies for the first 20 elements in the Periodic Table*

We can use the three ideas above to explain the change in ionisation energy with increasing proton number.

- Across a period there is a general increase in ΔH_{i1}. The increased nuclear charge (number of protons) outweighs the other two factors because the successive electrons added are being placed in the same outer electron shell. There is not much difference in shielding since there is the same number of inner electron shells.
- When a new period starts, there is a sharp decrease in ΔH_{i1}. The outer electron goes into a shell further from the nucleus and there is more shielding. Both these factors outweigh the effect of greater nuclear charge. This also explains why ΔH_{i1} decreases down a group.

Did you know?

The p_x, p_y and p_z orbitals in a particular sub-shell are equal in energy. The electrons are added one by one to these separate orbitals before they are paired up. This minimises repulsions between electrons. The electrons in the separate p orbitals of nitrogen and oxygen are shown here:

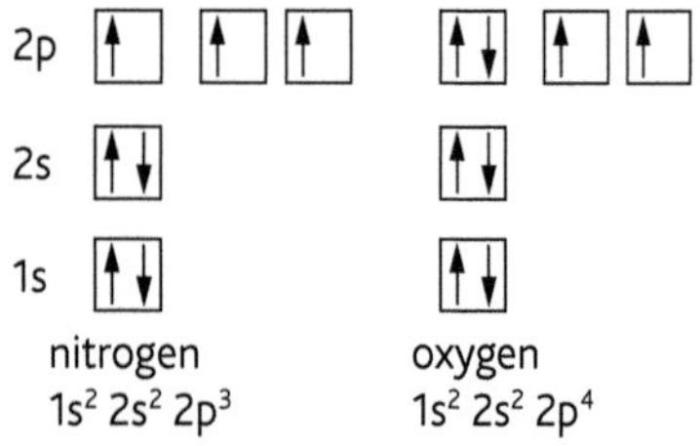

The extra electron in oxygen has to go into an orbital which is already half-filled. The extra repulsion of electrons in the same orbital is called spin-pair repulsion.

- The decrease in ΔH_{i1} from Be → B is due to the fact that the fifth electron in B goes into the 2p sub-shell, which is slightly further from the nucleus than the 2s sub-shell. The shielding also increases slightly. These two factors outweigh the effect of the increased nuclear charge. A similar reason explains the dip in ΔH_{i1} from Mg → Al.

$$Mg = 1s^2 2s^2 2p^6 3s^2 \qquad Al = 1s^2 2s^2 2p^6 3s^2 3p^1$$

- The decrease in ΔH_{i1} from N → O is due to spin-pair repulsion. The eighth electron in O goes into an orbital which already has an electron in it. The increased repulsion of the similarly charged particles makes it easier to remove the eighth electron.

Successive ionisation energies

We can remove electrons from an atom one by one. The energies associated with removing these electrons are called the successive ionisation energies. They are given the symbols ΔH_{i1}, ΔH_{i2}, ΔH_{i3}, etc. For example:

$$O^+(g) \rightarrow O^{2+}(g) + e^- \quad \Delta H_{i2} = +3390\,kJ\,mol^{-1}$$
$$O^{2+}(g) \rightarrow O^{3+}(g) + e^- \quad \Delta H_{i3} = +5320\,kJ\,mol^{-1}$$

Electron configuration from ionisation energies

We can work out the electron configuration of an element from successive ionisation energies. Figure 1.5.3 shows how this is done. In some instances it may also be possible to distinguish between s and p sub-shells by small changes in successive ionization energies.

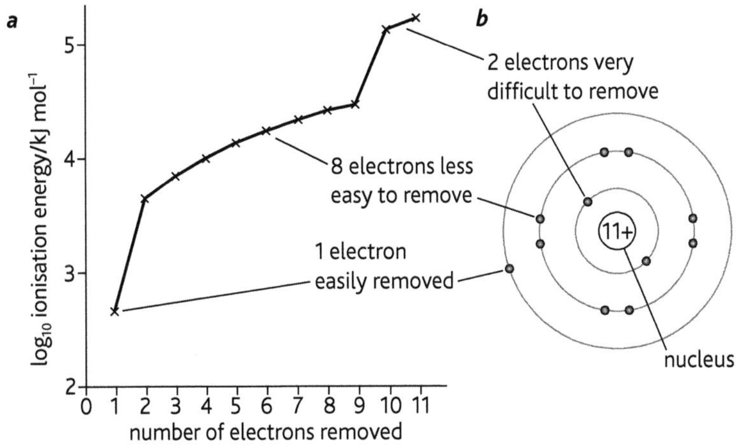

Figure 1.5.3 ***a*** *Successive ionisation energies for a sodium atom;* ***b*** *How the arrangement of electrons can deduced from these ionisation energies*

Exam tips

In considering ionisation energies, remember that:

1. The ions formed are ***always*** positive ions.
2. The atoms or ions are always gaseous.
3. The charge on the ion *formed* is the same as the number of the (successive) ionisation energy number e.g. for the fourth ionisation energy the ion formed is a 4+ ion.

Key points

- The first ionisation energy is the energy needed to remove a mole of electrons from a mole of gaseous atoms.
- The ionisation energies required to remove electrons one at a time from a mole of gaseous atoms are called the successive ionisation energies.
- Ionisation energy values depend on the nuclear charge, the distance of the outer electrons from the nucleus and shielding.
- Trends in ionisation energy can be explained using these three factors.
- The values of successive ionisation energies of an atom provide evidence for the electronic configuration.

Revision questions

1 State the four main tenets of Dalton's atomic theory and explain any modifications to his theory in light of later discoveries concerning the atom.

2 Explain the direction and degree of deflection of protons, electrons and neutrons in an electric field.

3 What is the rationale for using the relative masses and relative charges of sub-atomic particles rather than their absolute masses and charges?

4 How many protons, electrons and neutrons are in the following atoms?

a $^{27}_{13}Al$

b $^{39}_{19}K$

c $^{131}_{53}I$

d $^{239}_{94}Pu$

5 Calculate the relative atomic mass of boron given that its two isotopes ^{10}B and ^{11}B have relative abundances of 18.7% and 81.3% respectively.

6 a Write equations showing the α-decay of the following radioisotopes:

i $^{238}_{92}U$

ii $^{222}_{88}Ra$

b Write equations showing the β-decay of the following radioisotopes:

i $^{234}_{90}Th$

ii $^{14}_{6}C$

7 State the uses of four named radioisotopes.

8 Explain how an atomic emission spectrum is produced.

9 How is the

a Lyman series

b Balmer series

produced in the hydrogen emission spectrum?

10 Write the equation that indicates the energy difference between two energy levels and state the meaning of all of the symbols used.

11 State the maximum number of electrons that can be held in an s, p and d sub-shell.

12 a What is an atomic orbital?

b Draw a:

i 2s orbital

ii 2p orbital

13 Write the electronic configurations of the following atoms and ions:

a Ca

b Zn

c Cl^{-}

d Mn^{2+}

14 a Define the term the 'first ionisation energy'.

b Write an equation showing the first ionisation energy of calcium.

15 State the three factors that influence the value of the ionisation energy.

16 Explain the general trend of an increase in ionisation energy across a period.

17 Explain why there is a decrease in the first ionisation energy between:

a Mg and Al

b P and S

18 The following data show the first seven successive ionisation energies of an element X. Suggest which group of the Periodic Table X belongs to.

Ionisation energy number	**Enthalpy/kJ mol^{-1}**
1st	737.7
2nd	1450.7
3rd	7732.7
4th	10 542.5
5th	13 630
6th	18 020
7th	21 711

2 Structure and bonding

2.1 States of matter and forces of attraction

Learning outcomes

On completion of this section, you should be able to:

- understand the relationship between forces of attraction and states of matter
- know that single covalent bonds are formed by sharing a pair of electrons between the atoms
- understand covalent bonding in terms of overlap of atomic orbitals
- understand sigma and pi bonding.

States of matter and melting point

Substances with high melting points have strong forces of attraction between their atoms (or ions). Substances with low melting points have weak forces of attraction between their molecules.

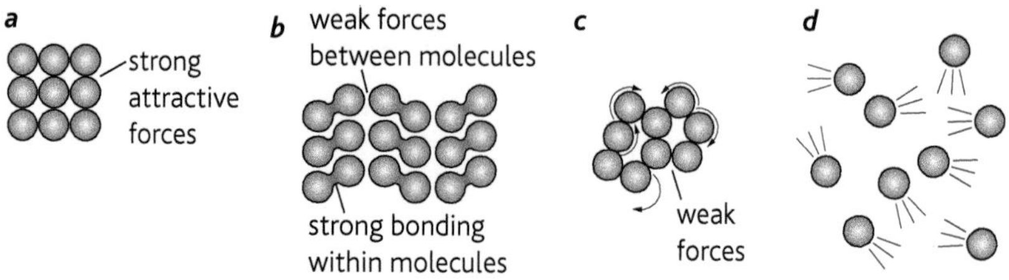

Figure 2.1.1 *The strength of forces and arrangement of particles in solids, liquids and gases*

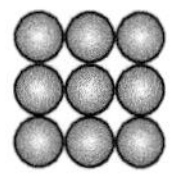

Giant structures with ionic or covalent bonds are solids with high melting points: It takes a lot of energy to break the many strong forces between the particles.

Particles are in fixed arrangement and do not move. They only vibrate.

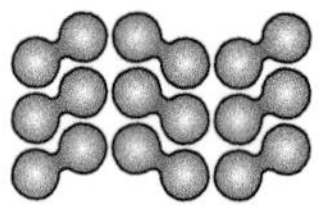

Molecular solids have low melting points: The forces within the molecules are strong. But the forces between the molecules are fairly weak so it does not take much energy to overcome these.

Particles are in fixed arrangement close together and do not move. They only vibrate.

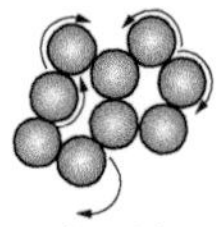

Liquids have low melting points. The forces within the molecules are strong. But the forces between the molecules are weak so it does not take much energy to overcome these.

Particles are close together but are more or less randomly arranged. They slide over each other.

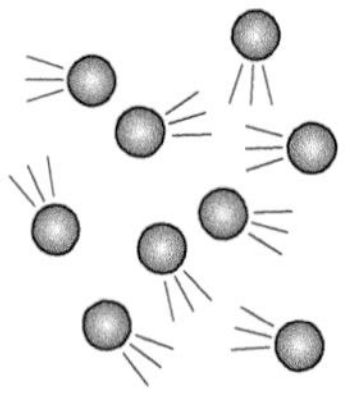

Gases have very low melting points and boiling points. The forces between the molecules are very weak so it does not take much energy to overcome these.

Particles are far apart and move randomly.

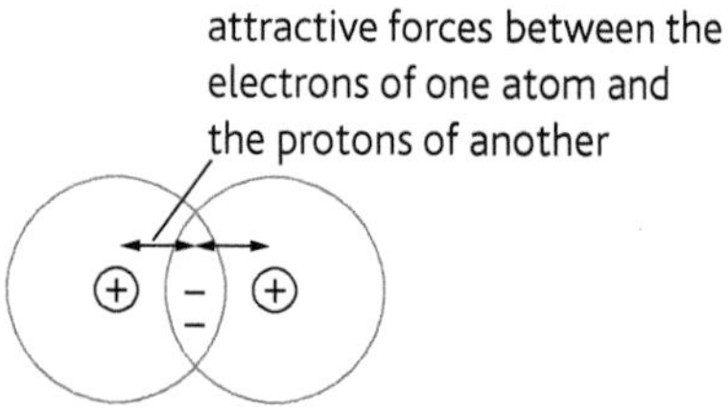

Figure 2.1.2
a *The formation of a covalent bond*

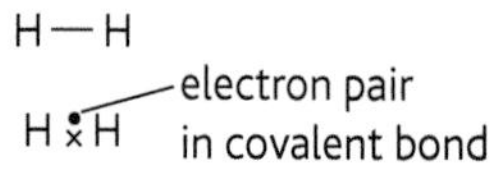

b *Two ways of showing the covalent bond in hydrogen*

Covalent bonding

A **covalent bond** is formed by the force of attraction between the nuclei of two neighbouring atoms and a pair of electrons between them. The attractive forces are in balance with the repulsive forces between the electron clouds when the nuclei are a certain distance apart. Covalent bonds are usually strong. It needs a lot of energy to break them.

Sigma bonds and pi bonds

A covalent bond is formed when atomic orbitals (Section 1.4) overlap. Each orbital which combines contributes one unpaired electron to the bond. The 'joined' orbital is called a **molecular orbital**. The greater the overlap of the atomic orbitals, the stronger is the covalent bond.

Sigma bonds (σ bonds)

Sigma bonds (σ bonds) are formed by the overlap of atomic orbitals along a line drawn between the two nuclei. The electron density of the bond formed is symmetrical about a line joining the two nuclei.

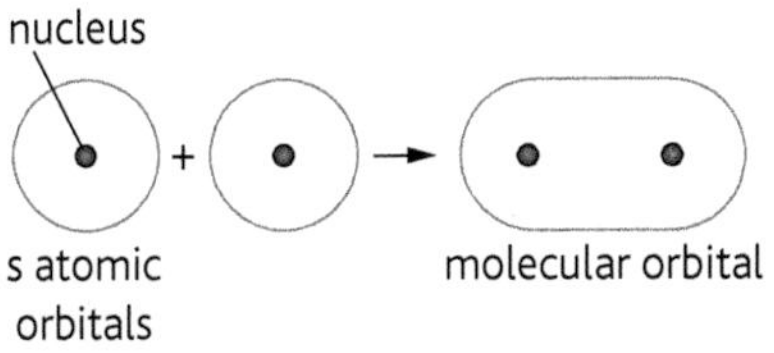

Figure 2.1.3 *Two s-type atomic orbitals overlap to form a molecular orbital*

When a p orbital combines with an s orbital, the molecular orbital is modified to include some s and some p character. The p orbital becomes slightly altered in shape so that, on combining to form a molecular orbital, one of the 'lobes' of the hour-glass shape becomes smaller. A similar thing happens when two p orbitals combine 'end-on'. In each case a σ bond is formed.

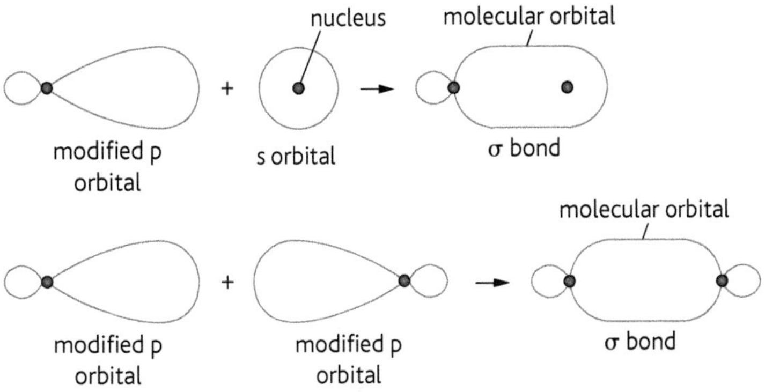

Figure 2.1.4 *The formation of sigma bonds by 'end-on' overlap of atomic orbitals*

Pi bonds (π bonds)

Pi bonds (π bonds) are formed by the sideways overlap of p atomic orbitals. The electron density of the bond formed is *not* symmetrical about a line joining the two nuclei. Figure 2.1.5 shows 2 p atomic orbitals overlapping to form a π bond (see also Section 2.9).

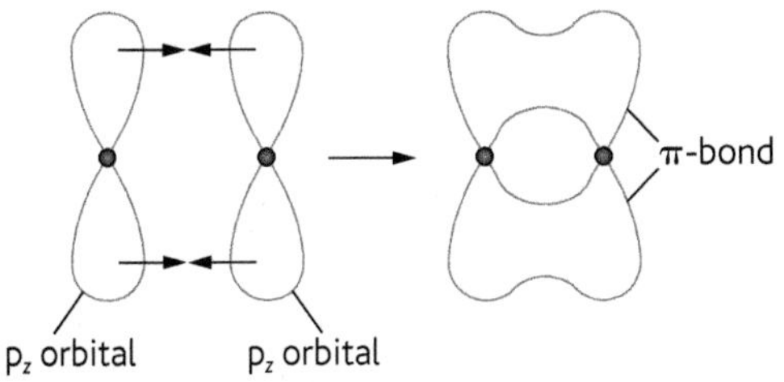

Figure 2.1.5 *The formation of a pi bond by 'sideways' overlap of p_z atomic orbitals*

Did you know?

The combining power of an atom is called its **valency** (from the Latin *valentia* – meaning powerful). So covalent means 'the power of (atoms) joining together'. The concept of valency is based on how many hydrogen atoms combine together with a given atom, e.g. the valency of oxygen is two because one oxygen can combine with two hydrogen atoms in water.

Exam tips

Although a π bond has two areas of electron density, you must remember that these two areas of electron density represent only one bond. Do not confuse it with a double bond, which contains one σ bond and one π bond.

Key points

- Solids with high melting points have strong forces of attraction between their atoms or ions.
- Liquids and gases have low melting points because of the weak forces between their molecules.
- A covalent bond is formed by the force of attraction between the nuclei of two atoms and the pair of electrons which forms the bond.
- Sigma bonds are formed by 'end-on' overlap of atomic orbitals.
- Pi bonds are formed by 'sideways' overlap of atomic orbitals.

Learning outcomes

On completion of this section, you should be able to:

- know that a single covalent bond is formed when two atoms share a pair of electrons
- draw dot and cross diagrams for a variety of compounds with single bonds only
- draw dot and cross diagrams for electron deficient compounds and compounds with an expanded octet of electrons.

Simple dot and cross diagrams

A shared pair of electrons in a single bond is represented by a single line.

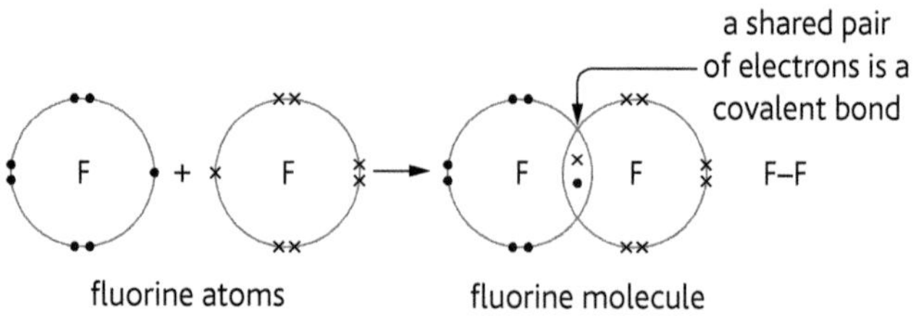

Figure 2.2.1 *The electron pairs in a fluorine molecule*

It is usually only the outer electrons which are used in covalent bonding. Electron pairs in the outer shell which are not used in bonding are called **lone pairs**.

Dot and cross diagrams can be used to show the arrangement of outer shell electrons in covalently bonded molecules. The main points are:

- Use a dot to represent the outer shell electrons from one atom and a cross to represent the outer shell electrons from another atom.
- Draw the outer electrons in pairs to emphasise the number of bond pairs and the number of lone pairs. It also reflects the fact that each orbital contains a maximum of two pairs of electrons.
- If possible, electrons are arranged so that each atom has four pairs of electrons around it (an octet of electrons/ the noble gas electron configuration). Hydrogen is an exception – it can only have two electrons around its nucleus when forming covalent bonds.
- When pairing electrons, a covalent bond is formed between an electron from one atom (dot) and an electron from another atom (cross).

Example: Methane

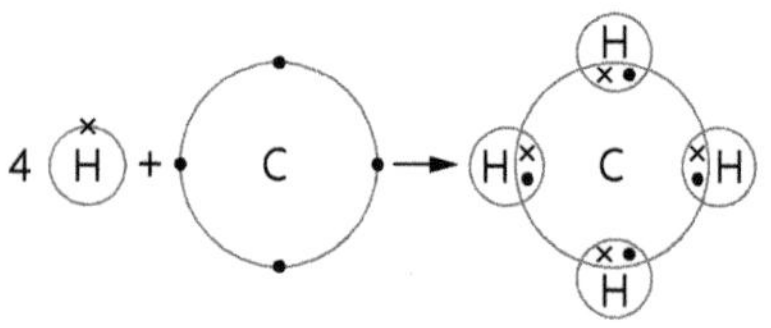

Figure 2.2.2 *Drawing a dot and cross diagram for methane*

Examples of dot and cross diagrams

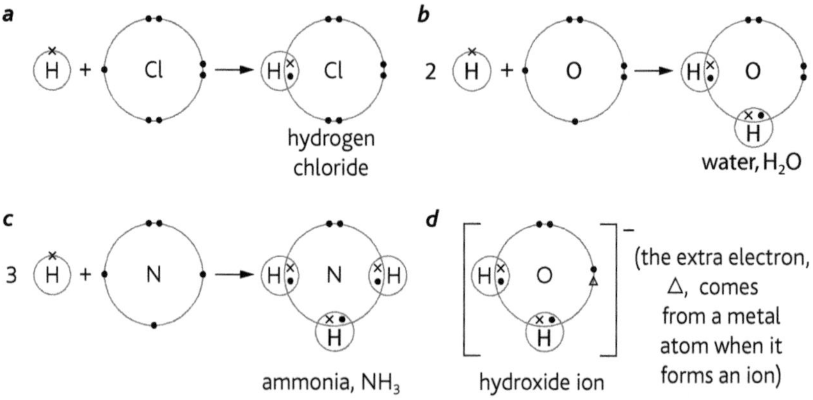

Figure 2.2.3 *Dot and cross diagrams for:* ***a*** *hydrogen chloride;* ***b*** *water;* ***c*** *ammonia;* ***d*** *the hydroxide ion*

Note that we can draw dot and cross structures for ions such as OH^- and NH_4^+ by applying the same rules. The square brackets around the ion show that the charge is spread evenly over the ion.

Did you know?

When we draw dot and cross diagrams, there is no difference in the electrons within each electron pair. The dots and crosses are just a book-keeping exercise to keep track of the electrons. The electron density in lone pairs of electrons is, however, greater than that of bonding pairs of electrons. For more information see Section 2.8.

Electron deficient molecules

Some molecules are unable to complete the octet of electrons when they form covalent bonds. These molecules are said to be **electron deficient**. An example is boron trichloride, BCl_3. This only has six electrons around the boron atom.

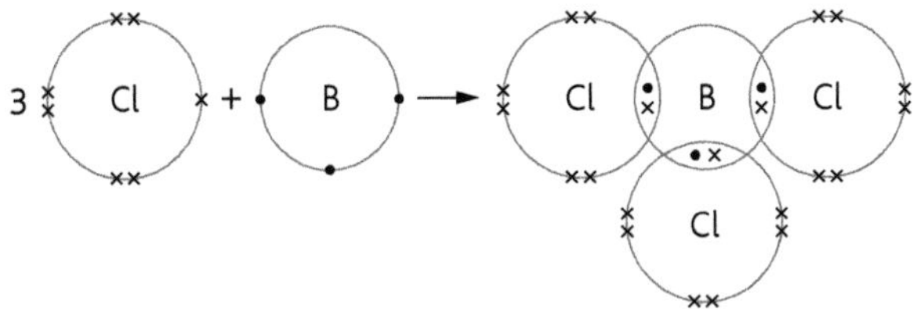

Figure 2.2.4 *Dot and cross diagrams for boron trichloride,* BCl_3

Molecules with an expanded octet

Some molecules can increase the number of electrons in their outer shell to more than 8. These molecules are said to have an **expanded octet** of electrons. An example is sulphur hexafluoride, SF_6. This has 12 electrons around the sulphur atom.

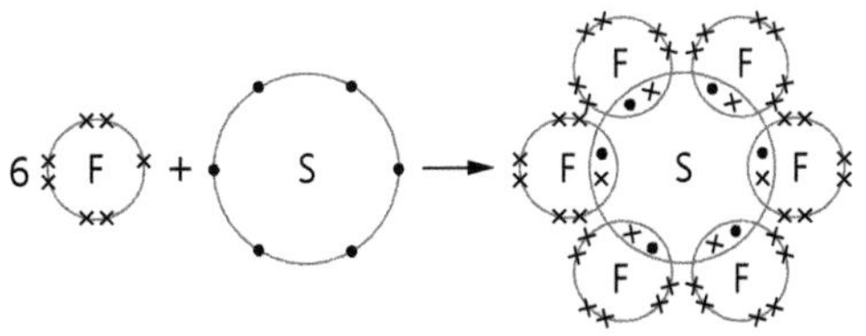

Figure 2.2.5 *Dot and cross diagrams for sulphur hexafluoride,* SF_6

Exam tips

It is a common error to try and write dot and cross diagrams for ionic compounds as if they were covalent molecules. If you have a compound of a metal and non-metal then it is likely to be ionic unless the metal ion is very small and highly charged compared with the non-metal ion. To see how to write dot and cross diagrams for ionic compounds see Section 2.4.

Key points

- A single covalent bond is formed when two atoms share a pair of electrons.
- When atoms form covalent bonds each atom usually has a full outer shell of electrons.
- Electron deficient atoms in a compound have fewer than 8 electrons in their outer shell.
- Atoms with an expanded octet have more than 8 electrons in their outer shell.
- Dot and cross diagrams show how the electrons pair together in a molecule or ion.

2.3 More dot and cross diagrams

Learning outcomes

On completion of this section, you should be able to:

- construct dot and cross diagrams for molecules with double and triple bonds
- describe co-ordinate bonding (dative covalent bonding)
- construct dot and cross diagrams involving co-ordinate bonds.

Molecules with multiple bonds

Some atoms form bonds by sharing two pairs of electrons. A double bond is formed. The double bond is shown by a double line, e.g. for oxygen, O=O.

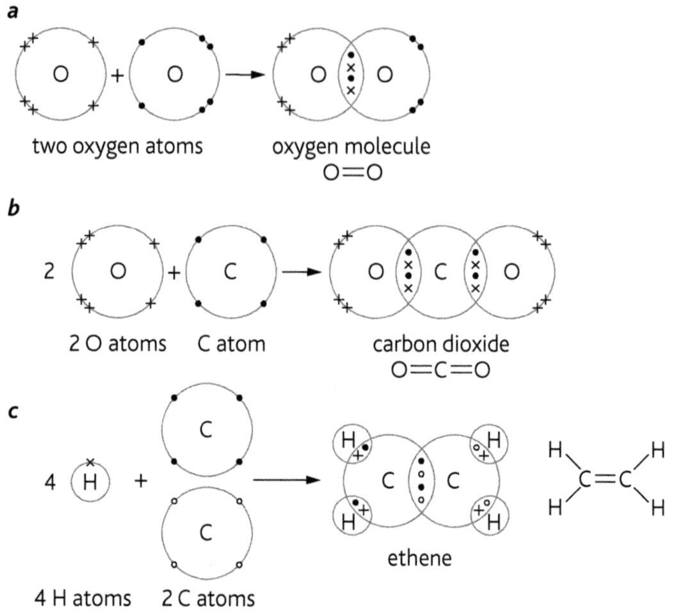

Figure 2.3.1 *Dot and cross diagrams for:* ***a*** *oxygen;* ***b*** *carbon dioxide;* ***c*** *ethene*

When atoms share three pairs of electrons, a triple bond is formed. The triple bond is shown by a triple line, e.g. for nitrogen N≡N.

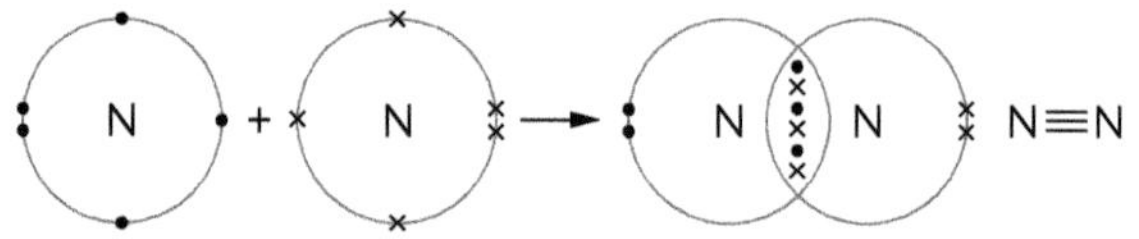

Figure 2.3.2 *A dot and cross diagram for nitrogen*

Co-ordinate bonding

A **co-ordinate bond (dative covalent bond)** is formed when one atom provides both the electrons for the covalent bond. For co-ordinate bonding to occur we need:

- one atom with a lone pair of electrons
- a second atom with an unfilled orbital.

The atom with the unfilled orbital (an electron deficient atom) accepts the lone pair of electrons to complete the outer shell of both atoms.

Example 1: Ammonium ion, NH_4^+

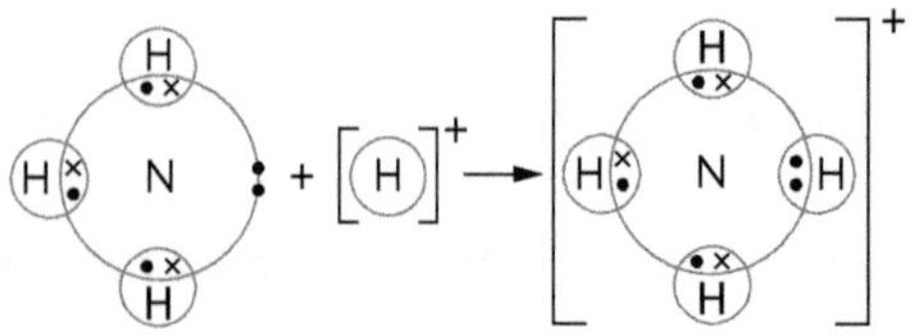

Figure 2.3.3 *A dot and cross diagram for an ammonium ion, NH_4^+*

In this example:

- The H^+ ion has space for two more electrons in its outer shell.
- The nitrogen atom has a lone pair of electrons.
- The lone pair on the nitrogen provides both electrons for the bond.
- Each atom now has a noble gas electron configuration. (H = 2; N = 2, 8).

The **displayed formula** (formula showing all atoms and bonds) shows the co-ordinate bond as an →. The head of the → points away from the atom which donates the lone pair.

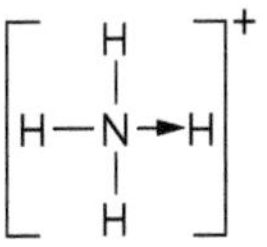

Figure 2.3.4 The displayed formula for an ammonium ion, NH_4^+

Example 2: Compound formed between BF_3 and NH_3

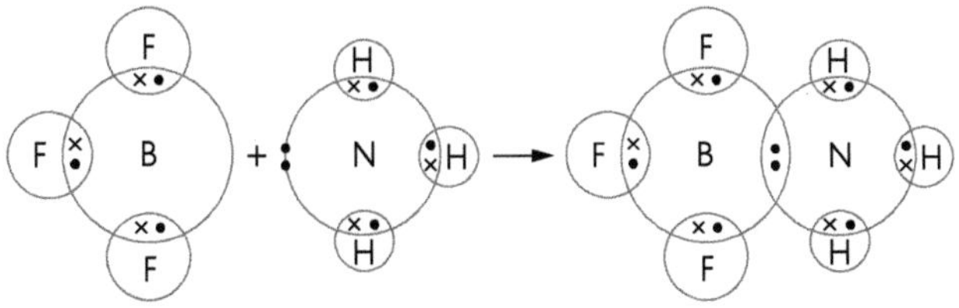

Figure 2.3.5 A dot and cross diagram for the co-ordination compound, BF_3NH_3

In this example:

- B has the simple electronic configuration 2, 6 when bonded to three fluorine atoms. There is still room for 2 more electrons to be added to its outer shell to form the nearest noble gas configuration.
- N has a lone pair of electrons.
- The nitrogen donates its lone pair to the unfilled orbital of boron.
- Both boron and nitrogen have the simple electronic configuration 2, 8.

Example 3: Aluminium chloride, $AlCl_3$

Aluminium chloride is an electron deficient molecule – the aluminium atom is 2 electrons short of the 8 electrons required for the nearest noble gas configuration.

At room temperature, aluminium chloride exists as Al_2Cl_6 molecules. This is because the lone pairs of electrons on two of the chlorine atoms can form co-ordinate bonds with the aluminium atoms.

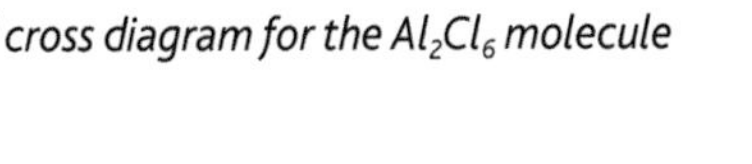

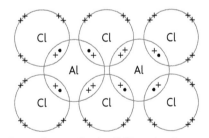

Figure 2.3.6 A dot and cross diagram for the Al_2Cl_6 molecule

Did you know?

Hydrated aluminium chloride, $AlCl_3{\cdot}6H_2O$ is an ionic compound, but anhydrous aluminium chloride is a molecule with the formula Al_2Cl_6 which has co-ordinate bonding between two of the Cl and Al atoms. The hydrated compound is ionic because the water molecules stabilise the Al^{3+} ions. They do this by co-ordinate bonding between water molecules and the highly charged Al^{3+} ion.

Key points

- A double bond contains two pairs of electrons and a triple bond contains three pairs of electrons.
- In co-ordinate (dative covalent) bonding one atom provides both electrons for the bond.

2.4 Ionic and metallic bonding

Learning outcomes

On completion of this section, you should be able to:

- know that ions are formed when atoms gain or lose electrons
- write dot and cross diagrams for ionic structures
- describe the structure of metals.

Exam tips

1 Remember that for metal atoms in Groups I to III, the charge on the ion formed is the same as the group number. For non-metal atoms in Groups V to VII, the charge on the ion is group number −8. For example, S forms the sulphide ion, S^{2-} (6 − 8 = −2).

2 Remember that some positive ions do not contain metals, e.g. H^+ and NH_4^+ ions.

The formation of ionic compounds

- When electrons are transferred from metal atoms to non-metal atoms, ions are formed.
- The outer shell of both ions formed have the noble gas electron configuration:

$$\underset{2,\,8,\,8,1}{K} \rightarrow \underset{[2,8,8]^+}{K^+} + e^- \qquad \underset{2,6}{O} + 2e^- \rightarrow \underset{[2,8]^{2-}}{O^{2-}}$$

- Positive ions are formed by loss of electrons and negative ions are formed by gain of electrons.
- In an ionic compound the number of positive and negative charges must balance, e.g. for calcium chloride the ions present are Ca^{2+} and Cl^-. So the formula for calcium chloride is $[Ca^{2+}]\ 2[Cl^-]$, or more simply, $CaCl_2$.

Dot and cross diagrams for ionic structures

When writing dot and cross diagrams for ionic structures we usually show only the outer electron shells because these are the ones involved in electron transfer. Figure 2.4.1 shows how to construct a dot and cross diagram for magnesium oxide. Note:

- The ions formed have a full outer shell of electrons (electron configuration of the nearest noble gas).
- The charge on the ion is placed at the top right.
- The square brackets indicate that the charge is spread throughout the ion.

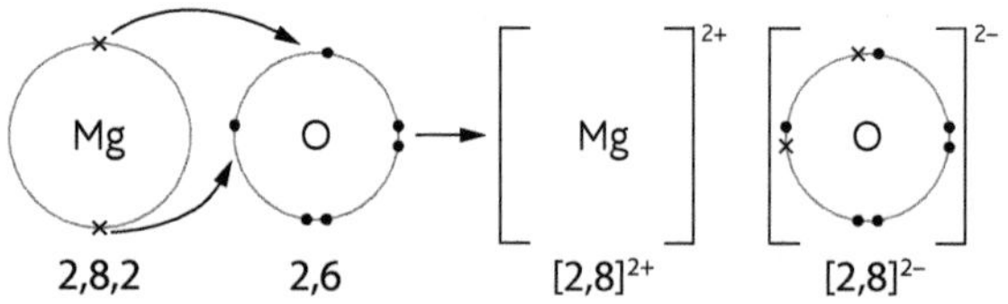

Figure 2.4.1 *Constructing a dot and cross diagram for magnesium oxide*

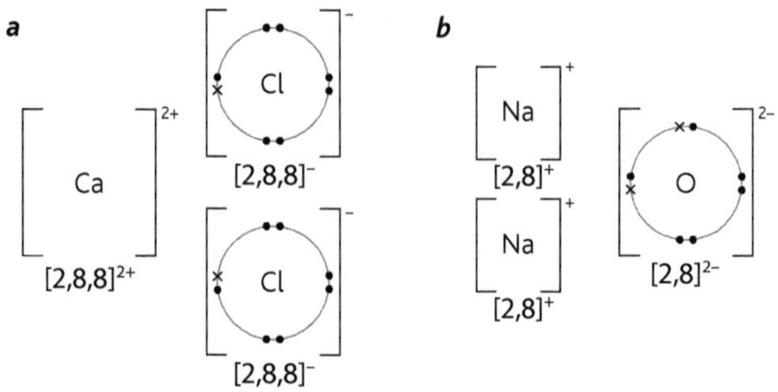

Figure 2.4.2 *Dot and cross diagrams for: **a** calcium chloride; **b** sodium oxide*

Ionic bonding

The **ionic bond** is an electrostatic force of attraction between oppositely charged ions. The net attractive forces between these ions results in the formation of a **giant ionic structure.** In this structure the ions are regularly arranged in a three-dimensional **lattice** (see also Section 2.6). The electrostatic attractive forces between the ions act in all directions and the bonding is very strong.

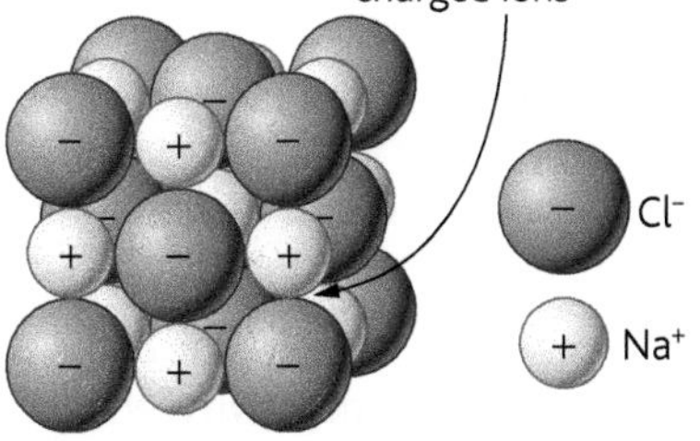

Figure 2.4.3 *Part of a giant ionic lattice of sodium chloride*

Metallic bonding

Most metals exist in a lattice of ions surrounded a 'sea' of **delocalised electrons**. Delocalised electrons are those which are not associated with any particular atom or ion. They are free to move between the metal ions.

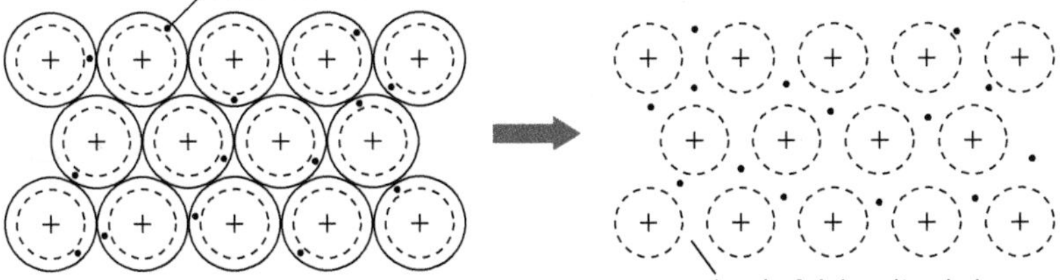

Figure 2.4.4 *The structure of a typical metal. (Note: the diagram shows only one layer of atoms for clarity. In reality there would be other layers on top of and beneath the layer shown.)*

In this structure:

- The number of delocalised electrons depends on the number of electrons lost by each metal atom.
- The positive charges are held together by their strong electrostatic attraction to the delocalised electrons.
- This strong electrostatic attraction acts in all directions.
- So metallic bonding is usually strong.

The strength of metallic bonding increases with:

- increasing positive charge on the ions
- decreasing size of the metal ions
- increasing number of delocalised electrons.

Exam tips

Remember that the big difference between metallic and ionic bonding is:

Ionic: the negative charges are ions.

Metallic: the negative charges are electrons.

Key points

- Ions are formed when atoms gain or lose electrons.
- Ions generally have a full outer shell of electrons (noble gas structure).
- Ionic bonding is the net attractive force between positively and negatively charged ions.
- Metallic bonding is a lattice of positive ions in a 'sea' of delocalised electrons.

2.5 Electronegativity and intermolecular forces

Learning outcomes

On completion of this section, you should be able to:

- understand the meaning of the terms 'electronegativity' and 'bond polarity'
- describe the three types of weak intermolecular forces: permanent dipole–dipole, van der Waals, hydrogen bonding.

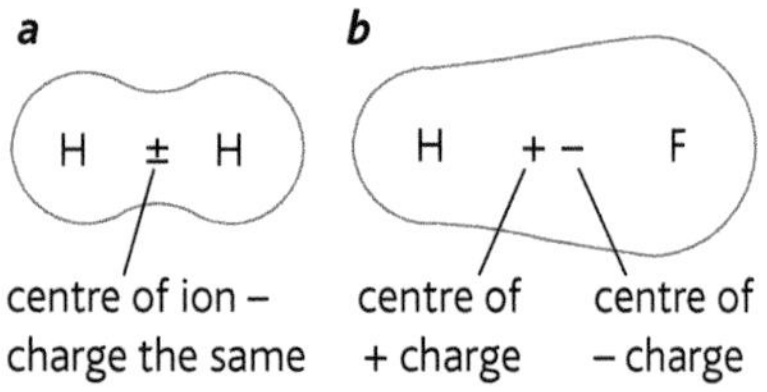

Figure 2.5.1 *a Hydrogen is a non-polar molecule because the centres of + and – charge coincide.* ***b*** *Hydrogen fluoride is polar because the centres of + and – charge do not coincide.*

Exam tips

We often talk about bonding in general terms as being strong or weak. Many chemists however reserve the term 'bonding' for stronger bonding, especially covalent bonding. When we are talking about weak bonding between molecules, it is more accurate to refer to them as forces between molecules. It may help to make the distinction between forces within molecules (strong bonding) and forces between molecules (weak).

Electronegativity

Electronegativity is the ability of a particular atom involved in covalent bond formation to attract the bonding pair of electrons to itself.

- Electronegativity increases across a period from Group I to Group VII.
- Electronegativity decreases down any group.
- The order of electronegativity is: F > O > N > Cl > Br...> C > H.

Polarity in molecules: bond polarisation

- If the electronegativity values of the two atoms in a covalent bond are the same, we say that the bond is **non-polar**.
- If the electronegativity values of the two atoms in a covalent bond are different, we say that the bond is **polar**.

We can also apply the idea of polarity to molecules:

- In a non-polar molecule the centres of positive and negative charge coincide.
- In a polar molecule the centres of positive and negative charge do not coincide. This means that one end of the molecule is slightly negatively charged, δ^-, and the other end is slightly positively charged, δ^+.

The degree of polarity is measured by a **dipole** moment. This is shown by the sign $\mapsto$. The $\rightarrow$ points to the partially negatively charged end of the molecule.

Examples:

$$\overset{\delta^+\quad\delta^-}{\overset{\longmapsto}{H - Cl}} \qquad \overset{\delta^-\quad\delta^+}{\overset{\longleftarrow\!\!+}{Cl - Br}}$$

If we know the shape of a molecule and the polarity of each bond, we can tell if the molecule as a whole is polar or non-polar. If the polarities of the bonds act so that they oppose each other, they may cancel each other out. The molecule is then non-polar (see Figure 2.5.2) because the centre of positive and negative charge is the same.

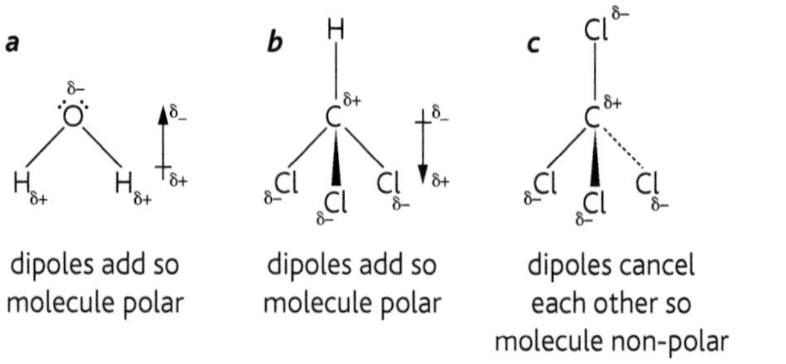

Figure 2.5.2 *The polarity in:* ***a*** *water;* ***b*** *trichloromethane;* ***c*** *tetrachloromethane*

Weak intermolecular forces

Intermolecular forces arise because of the attraction between the dipoles in neighbouring molecules. There are three types of intermolecular force:

- permanent dipole–dipole forces
- van der Waals forces
- hydrogen bonding.

Intermolecular forces are weak compared with covalent, ionic and metallic bonding. The comparative strength of these intermolecular forces are generally in the order: hydrogen bonding > permanent dipole–dipole > van der Waals.

Permanent dipole–dipole forces

Permanent dipole–dipole forces are the weak attractive forces between the δ^+ of the dipole of one molecule and the δ^- of the dipole of a neighbouring molecule (see Figure 2.5.3).

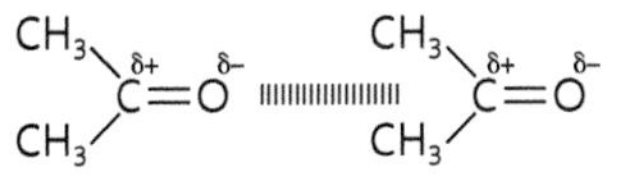

Figure 2.5.3 *Permanent dipole–dipole forces between two propanone molecules*

van der Waals forces

van der Waals forces of attraction are not permanent. All atoms and molecules, including noble gas atoms, have van der Waals forces.

- Electrons in atoms are always in motion.
- So the electron density may be greater in one part of the molecule than another.
- So an instantaneous dipole is formed.
- This dipole can induce the formation of a dipole in a neighbouring molecule.
- The two neighbouring molecules attract each other because of their dipoles.

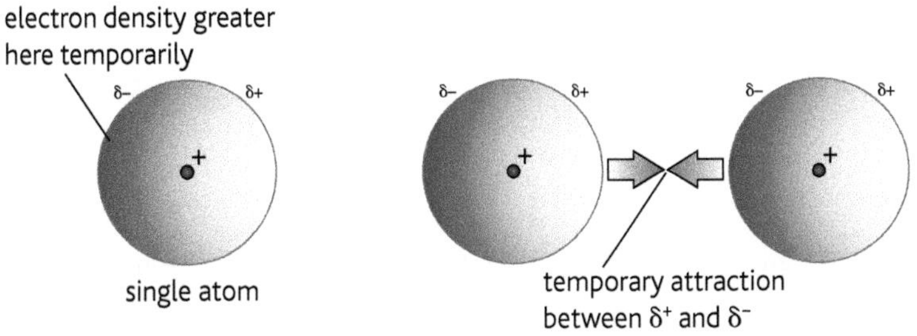

Figure 2.5.4 *van der Waals forces are only temporary because the electrons in the molecules of atoms are in constant movement*

Did you know?

The peculiar properties of water, e.g. lower density of ice than water and high surface tension, are due to more extensive hydrogen bonding than in many other molecules. Water molecules have two hydrogen atoms and two lone pairs of electrons. So water can form an average of two hydrogen bonds per molecule.

Hydrogen bonding

Hydrogen bonding is a special form of permanent dipole bonding. It requires:

- one molecule with an H atom covalently bonded to an F, O or N atom. These are the most electronegative atoms
- a second molecule having a F, O or N atom with a lone pair of electrons.

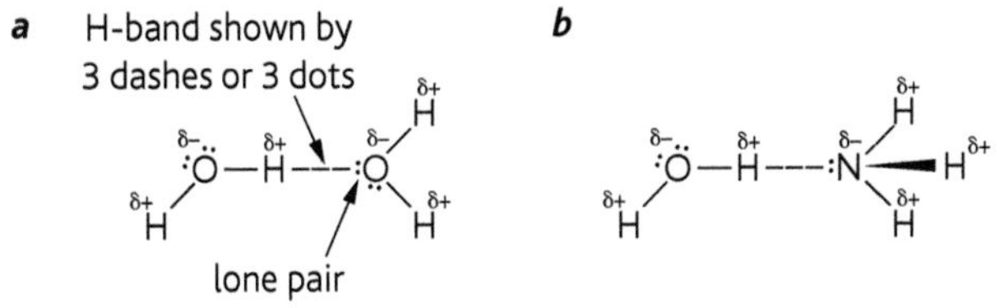

Figure 2.5.5 *Hydrogen bonding:* ***a*** *in pure water;* ***b*** *between water and ammonia molecules*

Key points

- Electronegativity is the ability of an atom in a covalent bond to attract electrons in the bond to itself.
- Electronegativity differences together with the structure of molecules can be used to determine whether a molecule is polar or non-polar.
- van der Waals forces are based on temporarily induced dipoles.
- Hydrogen bonding occurs between molecules having a lone pair of electrons on F, O or N and a hydrogen atom attached by a covalent bond to F, O or N.

2.6 Properties of giant structures

Learning outcomes

On completion of this section, you should be able to:

- describe giant structures in terms of their lattices
- understand how the properties of giant ionic structures, giant covalent structures and metals are related to their structure.

Lattices

A **lattice** is a regular three-dimensional arrangement of particles. Lattices with strong bonding between the particles are called **giant structures**. The three types of giant structures are:

- giant ionic, e.g. sodium chloride, magnesium oxide (see Section 2.4 for diagram)
- giant covalent (sometimes called giant molecular), e.g. diamond, silicon dioxide
- metallic, e.g. copper, iron (see Section 2.4 for diagram).

Properties of giant ionic structures

- Melting and boiling points are high: it takes a lot of energy to break the large number of strong bonds between the oppositely charged ions (see Section 2.4).
- Soluble in water: the ions can form ion–dipole bonds with water molecules.

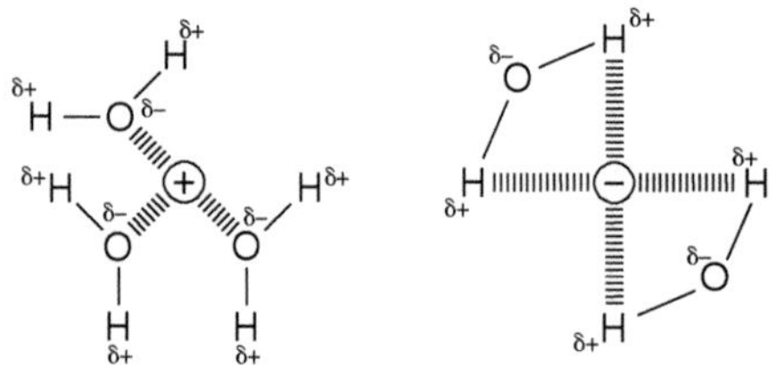

Figure 2.6.1 Water molecules forming ion–dipole bonds around + and – ions

- Conduct electricity only when molten or dissolved in water: the ions can only move when molten or dissolved in water. They cannot move in the solid.

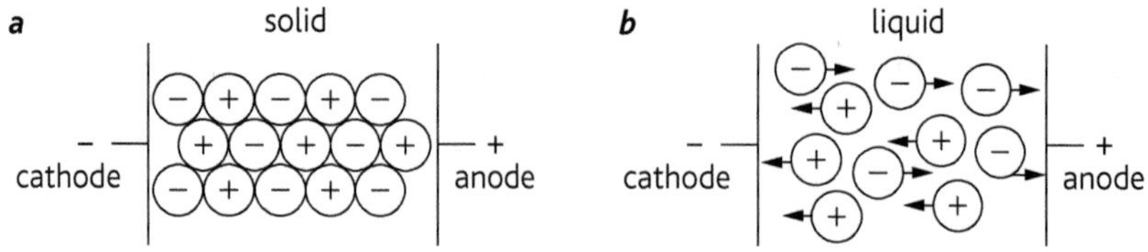

*Figure 2.6.2 **a** The ions cannot move in the solid state. **b** The ions are free to move in the liquid state.*

- Brittle: when a force is applied, the layers of ions slide. When many ions of the same charge are next to each other, there are many repulsive forces. So the lattice breaks easily.

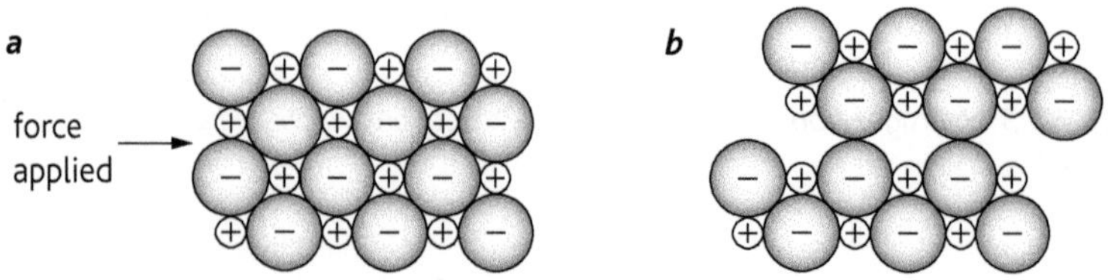

*Figure 2.6.3 **a** A small force is applied. **b** The ions move out of position and ions with the same charge are next to each other so the lattice breaks along this plane.*

Exam tips

A common error in exams is to suggest that electrons are involved when substances such as molten sodium chloride conduct electricity. First ask yourself if the structure is ionic (compound of reactive metal with non metal). If it is ionic, remember the 'ion' in ionic.

Properties of giant covalent structures

- Melting and boiling points are high: it takes a lot of energy to break the large number of strong covalent bonds in the network of atoms.

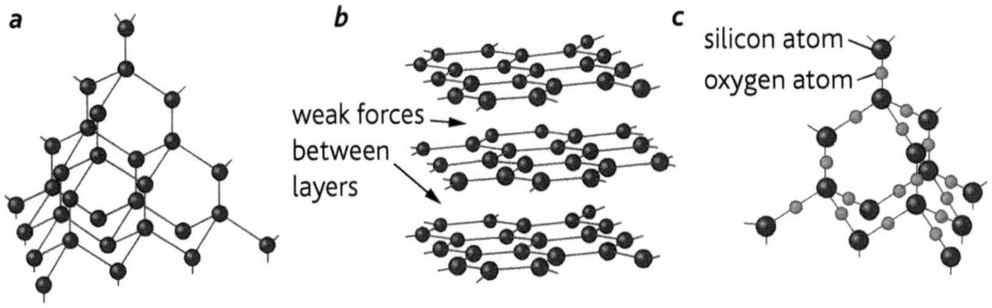

Figure 2.6.4 *The structures of: **a** diamond; **b** graphite; **c** silicon dioxide*

- Insoluble in water: the atoms are too strongly bonded to form bonds with water.
- Apart from graphite they do not conduct electricity: there are no electrons or ions that are free to move. Graphite conducts because it has some delocalised electrons. These electrons move anywhere along the layers when a voltage is applied.
- Apart from graphite they are hard: the three-dimensional network of strong bonds in different directions is too difficult to break. Graphite is soft because there are weak van der Waals forces between the layers. So the layers can slide over each other when a force is applied.

Did you know?

Each carbon atom in graphite is bonded to three others. This leaves a 'spare' electron in a p orbital in each carbon atom. These electrons form a huge delocalised system around all the rings in the graphite structure by π bonding. So the graphite conducts along the layers, but not between the layers.

Properties of metals

- Melting and boiling points are high: it takes a lot of energy to break the large number of strong forces of attraction between the ions and the delocalised electrons (see Section 2.4).
- Insoluble in water (although some react with water): the force of attraction between the ions and the delocalised electrons is too strong to form bonds with water. Some metals react with water because they lose electrons.
- Conduct electricity when solid or molten: the delocalised electrons are free to move when a voltage is applied.
- Malleable (can be beaten into shape) and ductile (can be drawn into wires): when a force is applied, the layers of metal ions can slide over each other. The attractive forces between the metal ions and the electrons can still act in all directions. So when the layers slide, new bonds can easily form. This leaves the metal in a different shape but still strong.

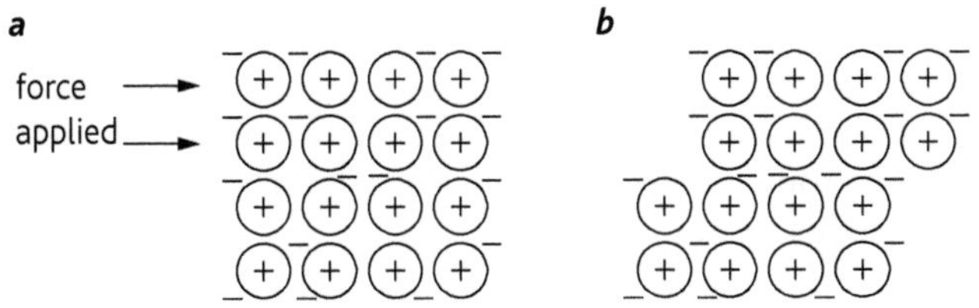

Figure 2.6.5 ***a*** *A small force is applied to a metal.* ***b*** *The metal ions slide out of position, but there are still strong forces between the ions and the delocalised electrons.*

Key points

- Giant ionic structures have a regularly repeating lattice of ions.
- Giant structures have high melting and boiling points because they have many strong bonds in many directions.
- Giant ionic structures are generally soluble in water. They only conduct electricity when molten or dissolved in water because the ions are free to move.
- Metals conduct electricity because some of the electrons are free to move throughout the structure.
- Giant covalent structures do not conduct electricity because they do not have free moving electrons or ions.
- Metallic structures are malleable and ductile because the layers of atoms can slide over each other.

2.7 Properties of simple molecular compounds

Learning outcomes

On completion of this section, you should be able to:

- understand how the properties of simple molecular compounds are related to their structures
- know that simple molecular solids often have a lattice structure
- understand how hydrogen bonding influences the properties of molecules
- describe the solubility of molecular compounds in polar and non-polar solvents.

Molecular lattices

Substances with a simple molecular structure such as iodine, I_2, can form crystals. This reflects a regular packing of the molecules in a lattice. The forces between the molecules are the weak van der Waals forces. So, simple molecular structures which are solids at room temperature are brittle. They do not conduct electricity because they have neither delocalised electrons nor ions.

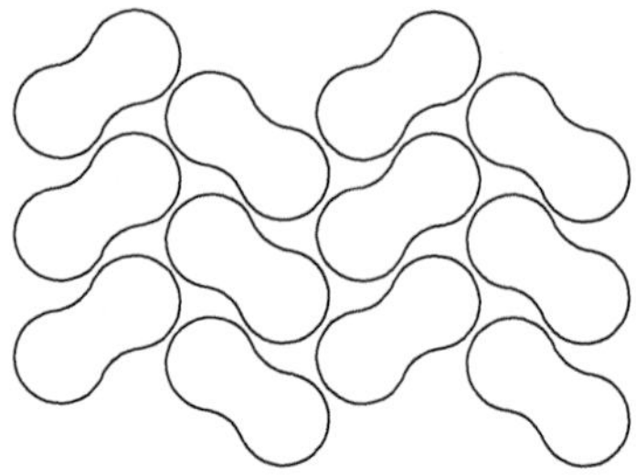

Figure 2.7.1 *The lattice structure of iodine (only shown in two dimensions)*

Melting and boiling points of simple molecular structures

Many simple molecular substances are liquids or gases. They have low melting and boiling points because it does not take much energy to overcome the weak intermolecular forces between the molecules. Solids with molecular lattices, such as iodine or sulphur, are also easily broken down when heated. It does not take much energy to overcome the weak van der Waals forces keeping the lattice together.

van der Waals forces increase with:

- increasing number of electrons in the molecule
- increasing number of contact points in the molecule (contact points are places where the molecules come close together).

So the melting points and boiling points of the noble gases and the halogens increase down each group as the number of electrons (and the relative mass) increases.

Pentane and 2,2-dimethylpropane have the same number of electrons, but the boiling point of pentane is higher. This is because there are more contact points for van der Waals forces to act over. The total van der Waals forces per molecule are therefore greater. So the boiling point is higher.

Large molecules, such as straight chained polymers, are solids at room temperature because they have very high relative molecular masses (large number of electrons). Also there are many contact points for the van der Waals forces to act over.

a

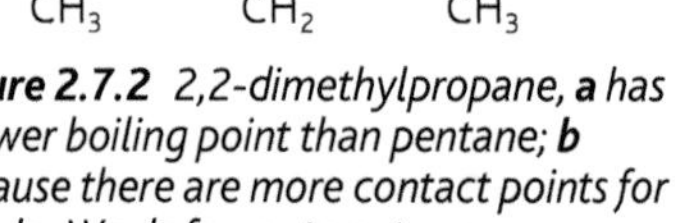

b

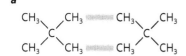

Figure 2.7.2 *2,2-dimethylpropane,* ***a*** *has a lower boiling point than pentane;* ***b*** *because there are more contact points for van der Waals forces to act over*

The anomalous properties of water

Water has a much higher boiling point than expected by comparison with other Group VI hydrides. The boiling points of the hydrides from H_2S to H_2Te increase gradually due to increased van der Waals forces. Water has a much higher boiling point than the other hydrides because it is extensively hydrogen-bonded. Since hydrogen-bonding is stronger compared with other intermolecular forces, it takes a lot more energy to vaporise water than the other Group VI hydrides.

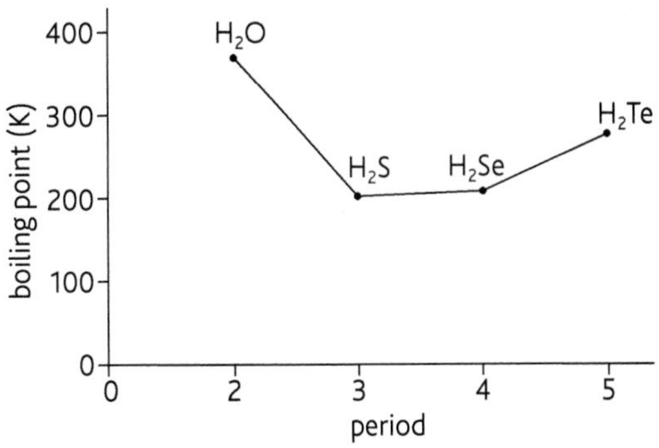

Figure 2.7.3 *The boiling points of the Group VI hydrides*

Most solids are denser than their corresponding liquids. Ice, however, is less dense than (liquid) water. This is because the molecules in ice are arranged in an 'open' lattice structure stabilised by hydrogen bonding. The molecules are not as close together as in liquid water.

Water also has a high surface tension due to hydrogen bonding.

Solubility of simple molecular substances

In order to dissolve, the strength of the attractive forces between solute and solvent is usually greater than between the solute molecules themselves. Most covalently-bonded molecules such as iodine, sulphur and butane, are insoluble in water. Many are non-polar and so water molecules are not attracted to them. They will, however, dissolve in non-polar solvents such as hexane if the new intermolecular forces formed between the solvent and solute are stronger than those between the solute molecules themselves. Some simple molecules are soluble in water, e.g. ethanol, C_2H_5OH and ammonia, NH_3. This is because they can form hydrogen bonds with water.

Figure 2.7.5 *Hydrogen bonding between:* ***a*** *ammonia and water;* ***b*** *ethanol and water*

Key points

- Simple molecular solids can form lattices with weak forces between the molecules.
- Simple molecular solids have low melting and boiling points because there are weak forces between the molecules.
- Hydrogen-bonded molecules have relatively higher melting and boiling points than expected and many are soluble in water.
- The total van der Waals forces between molecules may determine the state of a simple molecular compound at room temperature.

Exam tips

It is not always safe to assume that compounds with hydrogen bonding or dipole–dipole bonding always have a higher boiling point than those with van der Waals intermolecular forces. For example, the boiling point of SbH_3 (not H-bonded) is higher than that of NH_3 (H-bonded). This is because we are comparing molecules with very different masses, SbH_3 being more than 10× the mass of NH_3.

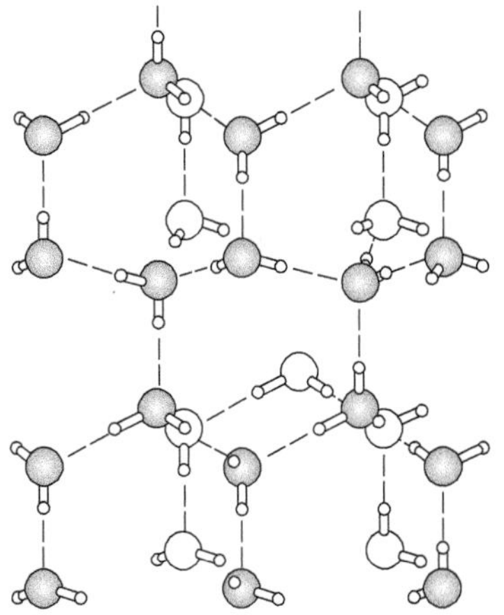

Figure 2.7.4 *The structure of ice*

Did you know?

You can float a needle on water because of the strong hydrogen bonding in the water. Place a needle on a piece of filter paper. Put the paper and needle on the surface of some very still water. When the filter paper sinks, the needle should remain on the surface of the water.

2.8 Shapes of molecules (1)

Learning outcomes

On completion of this section, you should be able to:

- predict the shapes and bond angles in simple molecules and ions.

Exam tips

Remember that 'tetra-' means four. So a tetrahedron has four faces.

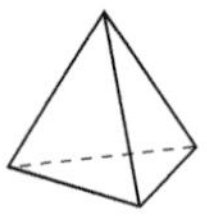

Figure 2.8.1 *A tetrahedron*

'Octa-' means eight. So an octahedron has eight faces.

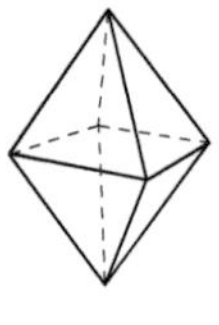

Figure 2.8.2 *An octahedron*

VSEPR theory

The **valence shell electron pair repulsion theory** (**VSEPR theory**) can be used to work out the shapes of molecules. It uses the following rules:

- Pairs of electrons in the outer shells of the atoms in a molecule repel each other and move as far apart as possible. This minimises repulsive forces in the molecule.
- Repulsion between lone-pairs and lone-pairs (of electrons) is greater than the repulsion between lone-pairs and bond-pairs of electrons.
- Repulsion between lone-pairs and bond-pairs (of electrons) is greater than the repulsion between bond-pairs and bond-pairs of electrons.

Shapes of molecules with only single bonds

The information below shows how VSEPR theory is used to work out the shapes of various molecules.

a

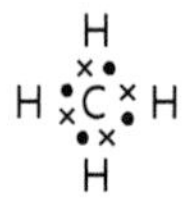

b

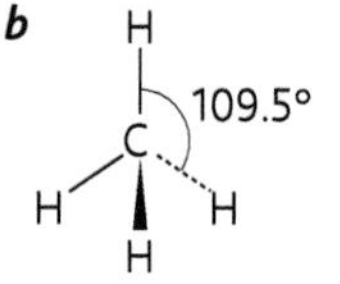

Four bond pairs. No lone pairs around C. Equal repulsion. So bond angles all 109.5°. Shape: **tetrahedral.** (NH_4^+ ion is also this shape)

c

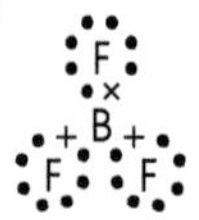

d

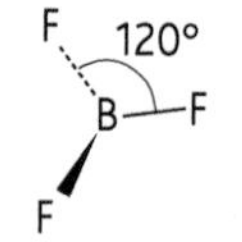

Three bond pairs. No lone pairs around B. Equal repulsion. So bond angles all 120°. Shape: **trigonal planar.**

e

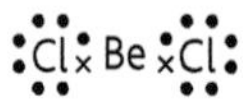

f

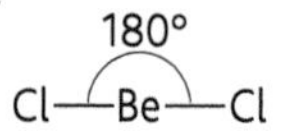

Two bond pairs. No lone pairs around Be. Equal repulsion. So bond angles both 180°. Shape: **linear.**

g

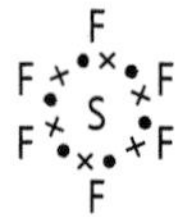

h

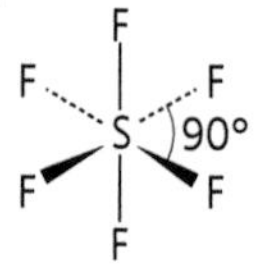

Six bond pairs. No lone pairs around S. Equal repulsion. So bond angles all 90°. Shape: **octahedral.**

i

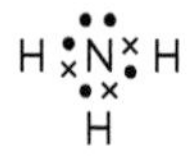

j

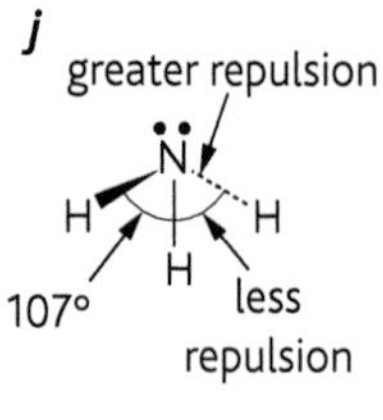

Three bond pairs. One lone pair around N. Greater repulsion between lone and bonding pairs. So bond angles close up to 107°. Shape: **pyramidal.**

k

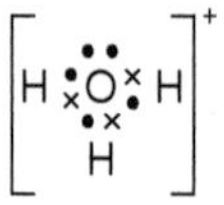

l

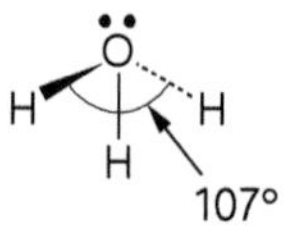

Three bond pairs. One lone pair around O. Greater repulsion between lone and bonding pairs. So bond angles close up to 107°. Shape: **pyramidal.** (The CH_3^- ion also has this shape.)

m

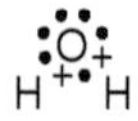

n

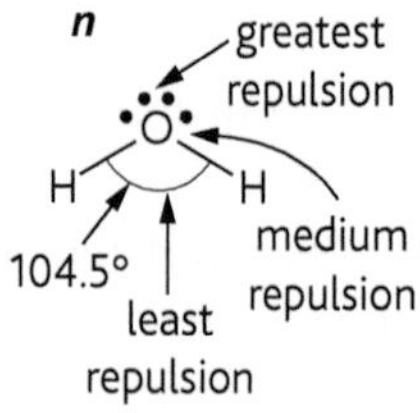

Two bond pairs. Two lone pairs around O. Greatest repulsion between lone pairs, less repulsion between lone pairs and bond pairs and least repulsion between bonding pairs. So bond angles close up to 104.5°. Shape: **non-linear, V-shaped.**

Shapes of molecules and compound ions with multiple bonds

The information below shows how VSEPR theory is used to work out the shapes of carbon dioxide as well as **oxo** ions (those containing oxygen and another non-metal).

a

b

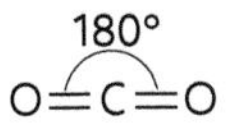

Two groups of two bonding pairs. No lone pairs. Equal repulsion. So bond angles both 180°. Shape: **linear.**

c

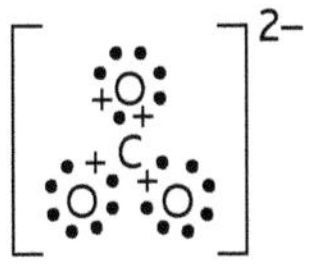

d

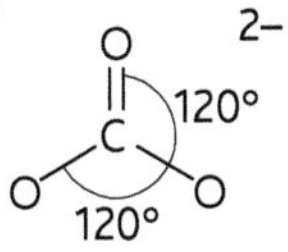

Three groups of bond pairs. No lone pairs. Equal repulsion. So bond angles 120°. Shape: **trigonal planar.** (The nitrate ion is also trigonal planar.)

e

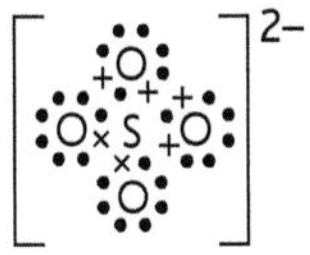

f

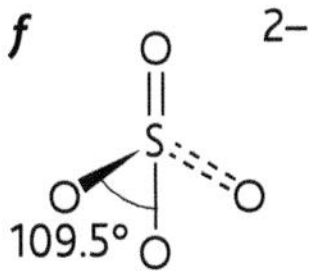

Four groups of bond pairs. No lone pairs. Equal repulsion. So bond angles 109.5°. Shape: **tetrahedral.**

Did you know?

Experimental data shows that all the carbon–oxygen bond lengths in CO_3^{2-} ions are the same. Similarly all the sulphur–oxygen bond lengths in the SO_4^{2-} ion are the same. This is due to hybridisation of the orbitals. Each bond has a bond length between that of a double and single bond between the relevant atoms.

Key points

- The shapes and bond angles in molecules can be predicted using VSEPR theory which depends on the number of lone pairs and bonding pairs of electrons around a particular atom.
- In VSEPR theory, lone-pair : lone-pair repulsion > lone-pair : bond-pair repulsion > bond-pair : bond-pair repulsion.

2.9 Shapes of molecules (2)

Learning outcomes

On completion of this section, you should be able to:

- predict the shapes and bond angles in simple organic compounds
- understand the shapes of molecules using the idea of hybridisation of atomic orbitals
- use ideas of hybridisation to describe the shape of benzene.

Hybridisation of orbitals

Methane has four C–H bonds of equal length. The four unfilled C atomic orbitals can be thought of as being mixed so that each has $\frac{1}{4}$s character and $\frac{3}{4}$p character. This process of mixing atomic orbitals is called **hybridisation**. These mixed orbitals are called sp^3 orbitals. Hybridisation allows a greater overlap of atomic orbitals when a molecular orbital is formed and also allows each bond to be the same. They are all σ bonds with equal repulsion between them.

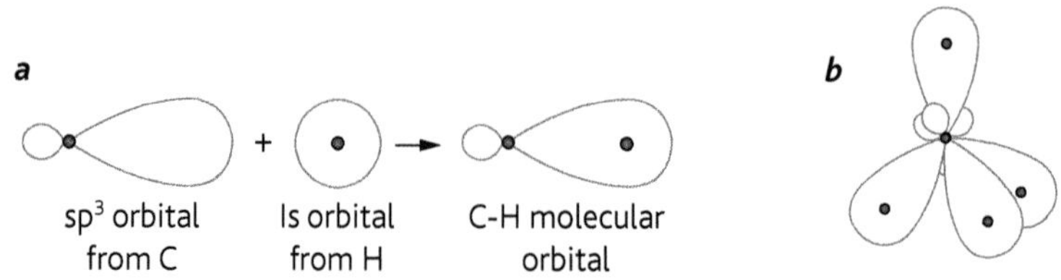

Figure 2.9.1 ***a*** *An sp^3 orbital from carbon combines with a 1s orbital from H to form a molecular orbital making up a C–H bond;* ***b*** *Methane has 4 directional hybrid orbitals*

Other hydrocarbons with single bonds also form molecular orbitals from hybridised sp^3 orbitals.

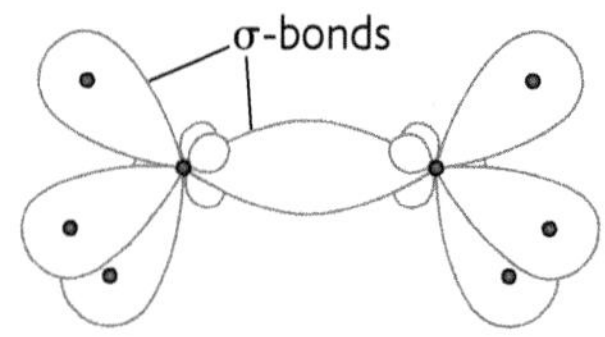

Figure 2.9.2 *The structure of ethane. The molecular orbitals formed from sp^3 hybrids allow each bond to be a σ-bond.*

The structure of ethene

The dot and cross diagram and shape of ethene are shown below.

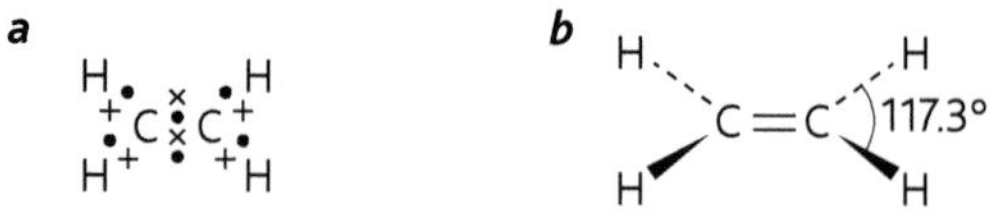

Figure 2.9.3 ***a*** *Dot and cross diagram for ethene;* ***b*** *Shape of an ethene molecule*

In ethene, one singly occupied 2s orbital and two of the three singly occupied 2p orbitals in each carbon atom hybridise to make three sp^2 orbitals. These have similar shapes to sp^3 orbitals. These sp^2 orbitals form σ bonds which are arranged in a plane making a bond angle of approximately 120° with each other since there is equal repulsion of the electrons.

The remaining 2p orbitals from each carbon atom overlap sideways to form a π bond.

The H–C–H bond angle in ethene is 117.3°, rather than the 120° expected, because of the influence of the π-bond whose electron density is at right angles to that of the plane of the carbon and hydrogen atoms. This bond angle allows the maximum overlap of the orbitals.

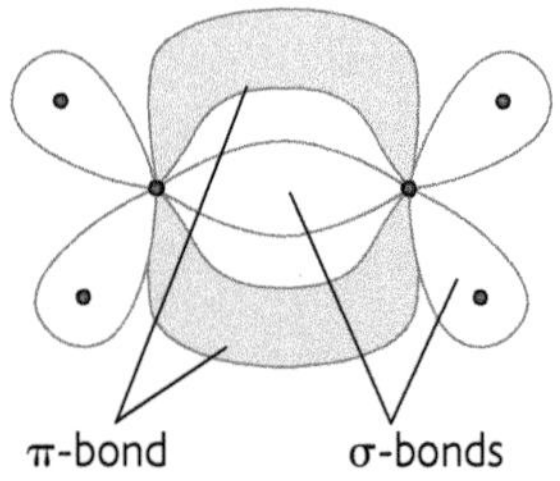

Figure 2.9.4 *Ethene has three sp^2 orbitals in one plane and a π-bond above and below this plane*

Resonance

In methane and ethene the electrons are localised, i.e. they are in particular positions. In some substances, the molecular orbitals extend over three or more atoms, allowing some of the electrons free movement over these atoms. These electrons are said to be **delocalised**.

Benzene, C_6H_6, has six carbon atoms arranged in a ring. Figure 2.9.5(a) shows a representation of benzene. The ↔ means that the actual structure is a single (composite) form which lies between these two structures. The bonds between the carbon atoms are neither double nor single bonds. They are somewhere in-between. The composite structure is called a **resonance hybrid**.

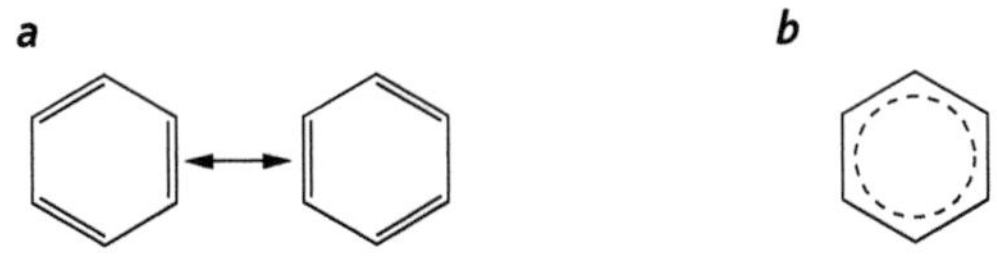

Figure 2.9.5 ***a*** *Two possible ways of representing benzene;* ***b*** *A modern representation of benzene*

In benzene, the six carbon atoms form a hexagon with three localised sp^2 hybrid orbitals (one to each hydrogen atom and two to other carbon atoms). The 3 sp^2 orbitals are arranged in a plane. So the bond angles are 120°. This leaves a single p orbital on each of the six carbon atoms. These orbitals overlap sideways to form a delocalised system of π bonds. The six electrons involved can move freely around the ring.

Figure 2.9.6 *The structure of benzene:* ***a*** *The p orbitals before overlap (the sp^2 hybrids are shown by the straight lines);* ***b*** *The delocalised system of π bonds*

In graphite, each carbon atom forms three localised sp^2 hybrid orbitals to the other carbon atoms. The remaining p orbitals overlap sideways on to form an extensive system of delocalised electrons. These electrons can move along the layers when a voltage is applied. This is why graphite conducts electricity.

Exam tips

It is a common error to think that resonance hybrids are mixtures of two or more forms of a structure. They are single structures. We can represent the 'in-between' structure in many cases by the use of dashed lines (see Figure 2.9.5(b))

Key points

- Hybridisation of s and p atomic orbitals results in the formation of an orbital with mixed character.
- The shapes of simple organic molecules can be predicted using VSEPR theory and the idea of hybridisation.
- Benzene is a resonance hybrid with a planar ring.

Revision questions

1 a Using the concept of atomic orbitals, explain how:
   i sigma and
   ii pi bonds are formed.
  b Which of these two bonds is stronger? Explain your answer.

2 Draw dot and cross diagrams for the following molecules:
  a H_2S
  b PCl_3
  c CO_2

3 a How is a co-ordinate bond formed?
  b H_3O^+ has a co-ordinate bond. Draw a dot and cross diagram to show the bonding in this ion.

4 Use dot and cross diagrams to illustrate the bonding in the following ionic compounds:
  a potassium sulphide
  b aluminium oxide

5 a An element has one electron in its outer shell and has filled inner shells. Describe the type of bonding in this element.
  b Would the melting point of this element be relatively high or low? Explain your answer.
  c Place the following Period 3 metals in order of increasing melting point:
     aluminium   sodium   magnesium

6 a Define 'electronegativity'.
  b What is the trend in electronegativity as you go down Group VII?
  c Explain your answer to part (b).

7 Identify the following molecules as polar or non-polar:
  a BF_3
  b CH_3Cl
  c CO_2
  d NH_3

8 a State the three types of intermolecular forces of attraction.
  b Identify the type of intermolecular force of attraction in the following molecules:
   i CO
   ii NH_3
   iii HF
   iv ICl

9 Magnesium oxide, diamond and aluminium are all solids. Using structure and bonding explain the differences, if any, among these three solids in terms of the following properties:
  a melting point
  b electrical conductivity
  c solubility in water.

10 Using structure and bonding explain why iodine:
  a is brittle
  b does not conduct electricity
  c iodine vaporises very easily.

11 Explain why water has a much higher boiling point than the other Group VI hydrides.

12 Which of the following molecules would have a higher melting point, BCl_3 or PCl_3? Explain your answer.

13 Using VSEPR theory, state and explain the shapes of the following molecules and ions:
  a H_3O^+
  b H_2S
  c BCl_3
  d CCl_4
  e PH_3

3 The mole concept

3.1 Equations and moles

Learning outcomes

On completion of this section, you should be able to:

- define the terms 'mole' and 'molar mass'
- construct molecular and ionic equations.

Exam tips

When balancing equations:

- Never alter the formula of a compound.
- Balance by putting numbers at the front of a formula.
- The first atom balanced should be the one which is easiest to balance in combustion reactions. Oxygen is often easiest to balance last.
- Numbers in front multiply all the way through the atoms in the formula.

Writing chemical equations

The four steps below show how to balance a simple chemical equation. The atom counting is shown by: C = ●, H = ○, O = x.

Step 1: Write down the formulae of all reactants and products:

$$CH_4(g) + O_2(g) \rightarrow CO_2(g) + H_2O(g)$$

Step 2: Count the number of atoms in each reactant and product:

$$CH_4 + O_2 \rightarrow CO_2 + H_2O$$

● ○○○○   xx   ● xx   ○○ x

Step 3: Balance one of the atoms e.g. hydrogen:

$$CH_4 + O_2 \rightarrow CO_2 + 2H_2O$$

● ○○○○   xx   ● xx   ○○○○ xx

Step 4: Keep balancing, one type of atom at a time until all atoms balance:

$$CH_4 + 2O_2 \rightarrow CO_2 + 2H_2O$$

● ○○○○   xxxx   ● xx   ○○○○ xx

You can instead just state the number of each type of atom on either side of the equation if you prefer.

The ratio of reactants and products in a balanced equation is called the **stoichiometry** of the reaction. In the equation for the combustion of methane above, the stoichiometry is 1 (methane): 2 (oxygen): 1 (carbon dioxide): 2 (water).

Ionic equations

Ionic compounds include salts such as ammonium sulphate and sodium carbonate as well as acids and alkalis. When ionic compounds dissolve in water, the ions separate. For example:

$$(NH_4)_2SO_4(aq) \rightarrow 2NH_4^+(aq) + SO_4^{2-}(aq)$$
$$H_2SO_4(aq) \rightarrow 2H^+(aq) + SO_4^{2-}(aq)$$
$$NaOH(aq) \rightarrow Na^+(aq) + OH^-(aq)$$

When ionic compounds react, only some of the ions take part in the reaction. The ions that play no part in the reaction are called **spectator ions**. To write an ionic equation:

Step 1: Write the full balanced chemical equation, e.g.

$$Mg(s) + CuSO_4(aq) \rightarrow MgSO_4(aq) + Cu(s)$$

Step 2: Write the charges on those substances which are ionic:

$$Mg(s) + Cu^{2+}(aq)\ SO_4^{2-}(aq) \rightarrow Mg^{2+}(aq)\ SO_4^{2-}(aq) + Cu(s)$$

Step 3: Cancel the spectator ions:

$$Mg(s) + Cu^{2+}(aq)\ \cancel{SO_4^{2-}}(aq) \rightarrow Mg^{2+}(aq)\ \cancel{SO_4^{2-}}(aq) + Cu(s)$$

Step 4: The ionic equation is that which remains:

$$Mg(s) + Cu^{2+}(aq) \rightarrow Mg^{2+}(aq) + Cu(s)$$

Exam tips

For precipitation reactions it is often easier to write an ionic equation from the ions making up the precipitate. For example, when aqueous lead nitrate is added to aqueous sodium chloride, lead chloride is precipitated. The ions which take part in the reaction are Pb^{2+} and Cl^-, so the equation is:

$$Pb^{2+}(aq) + 2Cl^-(aq) \rightarrow PbCl_2(s)$$

State symbols

State symbols are often added to equations to show the physical state of the reactants and products. These are placed after the formula of each reactant and product:

(s) solid (l) liquid (g) gas (aq) aqueous (a solution in water)

The mole and molar mass

The relative atomic mass or **relative molecular mass** of a substance in grams is called a **mole** of that substance (abbreviation mol). We use the ^{12}C scale as a standard in comparing masses accurately, so:

One mole is the amount of substance which has the same number of specific particles (atoms, ions or molecules) as there are atoms in exactly 12 g of the ^{12}C isotope.

The **molar mass**, *M*, is the mass of one mole of specified substance in grams. So the term 'molar mass' can apply to ionic compounds as well as molecules. For molecules we use the term 'relative molecular mass'. The units of *M* are $g\,mol^{-1}$.

The molar mass of a compound such as sodium sulphate, Na_2SO_4, is calculated by adding the relative atomic masses together taking into account the number of each type of atom:

$$A_r \text{ values: Na} = 23,\ \text{S} = 32;\ \text{O} = 16$$

$$M = \underset{2Na}{(2 \times 23)} + \underset{1S}{32} + \underset{4O}{(4 \times 16)} = 142\,g\,mol^{-1}$$

Exam tips

Accurate relative atomic masses are given in the Periodic Table to four significant figures, e.g. A_r[Na] = 22.99. You will often be given a rounded-up value for use in calculations, e.g. A_r[Na] = 23. Use this value when doing mole calculations. In doing simple calculations at this level in chemistry we take these 'rounded-up' values as being accurate enough to ignore any problems with significant figures (see Section 3.5).

Key points

- A mole is the amount of substance which has the same number of specified particles as there are atoms in exactly 12 g of the ^{12}C isotope.
- Molar mass is the mass of 1 mole of a substance in grams.
- Chemical equations show equal numbers of each type of atom in the reactants and products.
- In ionic equations the spectator ions are not shown.

3.2 Calculations using the mole concept

Learning outcomes

On completion of this section, you should be able to:

- perform calculations based on the mole concept.

Exam tips

When writing about moles, it is important to make clear what type of particles we are referring to. If we just say moles of chlorine, it is not clear whether its Cl atoms or Cl_2 molecules. For example in 35.5 g of Cl_2 there are 0.5 mol of Cl_2 molecules but 1.0 mol of Cl atoms.

Simple mole calculations

We can find the number of moles by using the relationship:

$$\text{number of moles (mol)} = \frac{\text{mass of substance (g)}}{\text{molar mass (g mol}^{-1})}$$

Worked example 1

How many moles of magnesium chloride, $MgCl_2$ are present in 19.1 g of magnesium chloride? (A_r values: Mg = 24.3, Cl = 35.5)

Step 1: Calculate the molar mass of $MgCl_2 = 24.3 + (2 \times 35.5)$
$= 95.3\,\text{g mol}^{-1}$

Step 2: Use the relationship: moles $= \dfrac{\text{mass (g)}}{\text{molar mass}} = \dfrac{19.1}{95.3} = 0.200\,\text{mol}$

Worked example 2

What mass of calcium hydroxide, $Ca(OH)_2$, is present in 0.025 mol of $Ca(OH)_2$?
(A_r values: Ca = 40.0, O = 16.0, H = 1.0)

Step 1: Calculate the molar mass of $Ca(OH)_2 = 40.0 + 2 \times (16.0 + 1.0)$
$= 74.0\,\text{g mol}^{-1}$

Step 2: Rearrange the equation in terms of mass:
mass (g) = moles × molar mass

Step 3: Substitute the values: mass (g) = 0.025 × 74.0 = 1.85 g

Reacting masses

To find the mass of products formed in a reaction we use:

- the mass of a specific reactant
- the molar mass of this reactant
- the balanced (stoichiometric) equation.

Worked example 3

Calculate the maximum mass of iron formed when 399 g of iron(III) oxide, Fe_2O_3, is reduced by excess carbon monoxide (A_r values: Fe = 55.8, O = 16.0).

Step 1: Calculate the relevant formula masses (Fe and Fe_2O_3):

$$Fe_2O_3 = (2 \times 55.8) + (3 \times 16.0) = 159.6$$

Step 2: Write the stoichiometric equation:

$$Fe_2O_3(s) + 3CO(g) \rightarrow 2Fe(s) + 3CO_2(g)$$

Step 3: Multiply each formula mass in grams by the relevant stoichiometric number:

$$Fe_2O_3 + 3CO \rightarrow 2Fe + 3CO_2$$
$$159.6\,\text{g} \rightarrow 2 \times 55.8\,\text{g}\ (111.6\,\text{g})$$

Step 4: Use simple proportion to calculate the amount of iron produced:

$$399\,\text{g} \rightarrow \frac{111.6}{159.6} \times 399 = 279\,\text{g Fe}$$

Worked example 4

We can do similar calculations to find the maximum mass of reactants used to obtain a given amount of product. The method shown above can be used, but a slightly different method is shown here.

Calculate the minimum mass of methane needed to produce 9.0 g of water when methane undergoes complete combustion (A_r values: C = 12.0, H = 1.0, O = 16.0).

Step 1: Calculate the relevant formula masses (CH_4 and H_2O):

$$CH_4 = 12.0 + (4 \times 1.0) = 16.0;\ H_2O = (2 \times 1.0) + (1 \times 16.0) = 18.0$$

Step 2: Calculate moles of water:

$$\text{moles} = \frac{\text{mass (g)}}{\text{molar mass}} = \frac{9.0}{18} = 0.50\,\text{mol}$$

Step 3: Use the stoichiometric equation to calculate the moles of methane:

$$\begin{array}{llll} CH_4(g) & + 2O_2(g) \rightarrow & CO_2(g) + & 2H_2O(g) \\ 1\,\text{mol} & \rightarrow & & 2\,\text{mol} \\ 0.25\,\text{mol} & \rightarrow & & 0.50\,\text{mol} \end{array}$$

Step 4: Calculate the mass of methane:

$$\begin{aligned} \text{mass (g)} &= \text{moles} \times \text{molar mass} \\ &= 0.25 \times 16 = 4.0\,\text{g} \end{aligned}$$

Ionic equations and mole calculations

The mole method can also be applied to masses of ions. The mass of ions are not significantly different from the mass of the atoms from which they are derived since the electrons have hardly any mass.

Worked example 5

Calculate the maximum mass of silver formed when 3.27 g of zinc reacts with excess aqueous silver ions (A_r values: Zn = 65.4, Ag = 108).

Step 1: Calculate the relevant formula masses (Zn and Ag^+):

$$Zn = 65.4;\ Ag^+ = 108$$

Step 2: Write the stoichiometric equation:

$$Zn(s) + 2Ag^+(aq) \rightarrow Zn^{2+}(aq) + 2Ag(s)$$

Step 3: Multiply each atomic mass in grams by the relevant stoichiometric number:

$$\begin{array}{lll} Zn(s) & + 2Ag^+(aq) \rightarrow Zn^{2+}(aq) + & 2Ag(s) \\ 65.4\text{ g} & \rightarrow & 2 \times 108\,\text{g (216 g)} \end{array}$$

Step 4: Use simple proportion to calculate the amount of silver produced:

$$3.27\,\text{g} \rightarrow \frac{3.27}{65.4} \times 216 = 10.8\,\text{g Ag}$$

Key points

- The mole concept can be used to calculate the masses of reactants required to form a specific mass of product(s).
- The mole concept can be used to calculate the mass of specific products formed from a given mass of reactants.

3.3 Empirical and molecular formulae

Learning outcomes

On completion of this section, you should be able to:

- deduce empirical formulae using absolute masses or relative masses of elements in a compound
- deduce molecular formulae.

Empirical formula and molecular formula

An **empirical formula** of a compound shows the simplest whole number ratio of the elements present in a molecule or formula unit of the compound.

A **molecular formula** shows the total number of each type of atom present in a molecule.

- The formula of an ionic compound is always its empirical formula.
- Many simple inorganic molecules have the same empirical and molecular formulae.
- Many organic molecules have different empirical and molecular formulae.

Some empirical and molecular formulae are shown below.

Compound	Empirical formula	Molecular formula
hydrogen peroxide	HO	H_2O_2
butane	C_2H_5	C_4H_{10}
benzene	CH	C_6H_6
sulphur dioxide	SO_2	SO_2
cyclohexane	CH_2	C_6H_{12}

Deducing an empirical formula

Worked example 1

Calculate the empirical formula of a compound of tin and chlorine which contains 5.95 g tin and 7.1 g chlorine (A_r values: Sn = 119, Cl = 35.5).

Step 1: Calculate the number of moles of each element.

$$\text{Sn} \qquad \frac{5.95}{119} = 0.05\,\text{mol} \qquad\qquad \text{Cl} \qquad \frac{7.1}{35.5} = 0.2\,\text{mol}$$

Step 2: Divide each by the lower number of moles

$$\frac{0.05}{0.5} = 0.1 \qquad \frac{0.2}{0.5} = 0.4$$

Step 3: Write the formula showing the simplest ratio: $SnCl_4$

Worked example 2

In this example, we are given information about % by mass rather than mass.

Calculate the empirical formula of a compound of carbon, hydrogen and iodine which contains 8.45% carbon, 2.11% hydrogen and 89.44% iodine by mass (A_r values: C = 12.0, H = 1.0, I = 127.0).

Step 1: Divide % by A_r

C	H	I
$\frac{8.45}{12.0} = 0.704$	$\frac{2.11}{1.0} = 2.11$	$\frac{89.4}{127.0} = 0.704$

Step 2: Divide by lowest number

$$\frac{0.704}{0.704} = 1 \qquad \frac{2.11}{0.704} = 3 \qquad \frac{0.704}{0.704} = 1$$

Step 3: Write the formula showing the simplest ratio: CH_3I

Exam tips

1 After Step 2 you may be left with figures that are not whole numbers, e.g. 1 P to 2.5 O. In this case you must multiply up to get a whole number (in this case by 2). So the formula is P_2O_5.

2 The empirical formulae of many compounds, especially inorganic ones, are fairly simple ratios. After Step 2 you may be left with a ratio such as 1 P to 1.46 O. It is therefore advisable to round up the 1.46 to 1.5. This will give an empirical formula of P_2O_3.

Deducing the molecular formula

The molecular formula is a simple multiple of the empirical formula. For example, the molecular formula of butane C_4H_{10} has twice the number of each type of atom as the empirical formula, C_2H_5. To deduce the molecular formula we need to know:

- the empirical formula
- the molar mass of the compound
- the empirical formula mass.

Worked example 3

A compound has an empirical formula CH_2 and a molar mass of 84. Deduce its empirical formula (A_r values: C = 12.0, H = 1.0).

Step 1: Find the empirical formula mass: $12.0 + (2 \times 1.0) = 14.0$

Step 2: Divide the molar mass of the compound by the empirical formula mass:

$$\frac{84}{14} = 6$$

Step 3: Multiply each atom in the empirical formula by the number deduced in Step 2:

$$CH_2 \times 6 = C_6H_{12}$$

Key points

- An empirical formula shows the simplest whole number ratio of atoms in a compound.
- Empirical formulae are deduced using masses or relative masses of the elements present in a compound.
- Molecular formulae show the total number of atoms present in a formula unit of a compound.
- A molecular formula can be deduced from the empirical formula if the relative molecular mass of the compound is known.

3.4 Using Avogadro's law

Learning outcomes

On completion of this section, you should be able to:

- apply Avogadro's law to deduce the stoichiometry of a reaction involving gases
- deduce molecular formulae from combustion data.

Did you know?

The number of atoms in a mole of atoms is very large: 6.02×10^{23} atoms mol^{-1}. Although this value is called the Avogadro constant, the number was first deduced by Johann Loschmidt. That is why the symbol for the Avogadro constant is *L*.

Exam tips

You may find it easier to do Worked example 2 by calculating the mass of CO_2 formed from 7.50 g of ethane, i.e.

30 g ethane produces
$2 \times 44\,g\ CO_2 = 88\,g\ CO_2$

So 0.25 g ethane produces $88 \times 0.25 = 22\,g\ CO_2$

$22\,g\ CO_2 = 22/44 = 0.5\,mol\ CO_2$

Avogadro's law

Avogadro's law states that *equal volumes of all gases at the same temperature and pressure have equal numbers of molecules.*

At room temperature and pressure (r.t.p.) 1 mole of any gas occupies $24.0\,dm^3$. At standard temperature and pressure (s.t.p.) it occupies $22.4\,dm^3$.

Applying Avogadro's law to the synthesis of water from hydrogen and oxygen:

$2H_2(g)$	+	$O_2(g)$	$\rightarrow$	$2H_2O(g)$
2 mol		1 mol		2 mol
2 volumes		1 volume		2 volumes
$48\,dm^3$		$24\,dm^3$		$48\,dm^3$ (at r.t.p.)

We can use Avogadro's law in mole calculations because if there are equal numbers of molecules in the same volume of gas, there are also equal numbers of moles.

Worked example 1

Calculate the mass of ethane ($M = 30\,g\,mol^{-1}$) in $240\,cm^3$ of ethane gas at r.t.p.

Step 1: Change cm^3 to dm^3: $240\,cm^3 \div 1000 = 0.240\,dm^3$

Step 2: Calculate the number of moles using: $\text{moles} = \frac{\text{volume (dm}^3)}{24}$

$$= \frac{0.240}{24} = 0.01\,mol$$

Step 3: Calculate mass: mass (g) = moles × M = $0.01 \times 30 = 0.3\,g$

Worked example 2

Calculate the volume of carbon dioxide formed at r.t.p. when 7.50 g of ethane is completely burnt in excess oxygen (A_r values: C = 12.0, H = 1.0).

Step 1: Calculate the number of moles of ethane:

$$\frac{7.50}{(2 \times 12.0) + (6 \times 1.0)} = 0.25\,mol$$

Step 2: Write the stoichiometric equation for the reaction and identify the relevant mole ratios:

$$2C_2H_6(g) + 7O_2(g) \rightarrow 4CO_2(g) + 6H_2O(l)$$
$$2\,mol \qquad\qquad\qquad 4\,mol$$

Step 3: Calculate the number of moles: 0.25 mol ethane $\rightarrow \frac{4}{2} \times 0.25$ mol CO_2

CO_2 formed from 0.25 mol ethane = 0.5 mol CO_2

Step 4: Calculate the volume of CO_2 at r.t.p: $0.5 \times 24 = 12\,dm^3$

Deducing the stoichiometry of a reaction

The worked example below shows how we can use Avogadro's law to deduce the stoichiometry of a reaction.

Worked example 3

A mixture of $40\,cm^3$ of hydrogen and $20\,cm^3$ of oxygen is reacted together. At the end of the reaction there is only water present. Use Avogadro's law to deduce the stoichiometry of the reaction.

Step 1: Write the unbalanced equation: $H_2(g) \;+\; O_2(g) \;\rightarrow\; H_2O(l)$

Step 2: Apply Avogadro's law:

	H_2	O_2
	$40\,cm^3$	$20\,cm^3$
	2 volumes	1 volume
	2 mole(cule)s	1 mole(cule)

Step 3: Write the equation: $2H_2(g) \;+\; O_2(g) \;\rightarrow\; 2H_2O(l)$

Deducing a molecular formula

The worked example below shows how we can use Avogadro's law to deduce the molecular formula of a compound using combustion data.

Worked example 4

Propane contains carbon and hydrogen only. When $25\,cm^3$ of propane reacts with exactly $125\,cm^3$ oxygen, $75\,cm^3$ of carbon dioxide is formed. Deduce the molecular formula of propane and write a balanced equation for the reaction.

Step 1: Write the information below the unbalanced equation:

$$\underset{25\,cm^3}{C_xH_y(g)} + \underset{125\,cm^3}{O_2(g)} \rightarrow \underset{75\,cm^3}{xCO_2(g)} + yH_2O(l)$$

Step 2: Find the simplest ratio of gases and use Avogadro's law:

$$\underset{\text{1 volume}}{C_xH_y(g)} + \underset{\text{5 volumes}}{5O_2(g)} \rightarrow \underset{\text{3 volumes}}{3CO_2(g)} + yH_2O(l)$$

Step 3: Deduce number of C atoms: $1\,mol\; C_xH_y \rightarrow 3\,mol\; CO_2$ (so x must be 3)

Step 4: Deduce the number of H atoms: $C_3H_y(g) + 5O_2(g) \rightarrow 3CO_2(g) + yH_2O(l)$

- 6 of the 10 oxygen atoms react with carbon
- so 4 oxygen atoms must react with hydrogen to form water
- so 4 moles of water are formed containing 8 hydrogen atoms which come from the propane.

Step 5: Write the balanced equation: $C_3H_8(g) + 5O_2(g) \rightarrow 3CO_2(g) + 4H_2O(l)$

Key points

- Avogadro's law states that equal volumes of all gases at the same temperature and pressure contain the same number of molecules.
- The stoichiometry of a reaction can be deduced by applying Avogadro's law.
- The molecular formula of a simple molecular compound can be deduced from combustion data by applying Avogadro's law.

3.5 Solution concentration and titrations

Learning outcomes

On completion of this section, you should be able to:

- understand how the mole concept can be used to determine the concentration of solutions
- know how to carry out acid–base titrations and how to process the results
- understand the use of significant figures in chemical calculations.

Exam tips

When dealing with significant figures (s.f.):

- zeros before numbers are not s.f., e.g. 0.036 has 2 s.f.
- Zeros after a figure are s.f., e.g. 0.0400 has 3 s.f.
- When doing a calculation with several steps, do not round up between steps. Only round up for your final answer.
- The final answer should be expressed to the same number of s.f. as the lowest number in the data provided.

The concentration of a solution

This refers to the amount of solute dissolved in a solvent to make a solution. In chemistry the units of solution concentration are usually expressed in moles per cubic decimetre ($mol\,dm^{-3}$). This is the **molar concentration.**

$$\text{concentration of solution (mol dm}^{-3}) = \frac{\text{number of moles of solute (mol)}}{\text{volume of solution (dm}^3)}$$

In calculations involving solution concentration in $mol\,dm^{-3}$ remember:

- to change mass in grams to moles
- to change cm^3 to dm^3 by dividing volume in cm^3 by 1000
- that moles solute = concentration × volume
- that volume of solution $= \dfrac{\text{moles of solute}}{\text{concentration}}$

Simple calculations involving concentration

Worked example 1

Calculate the concentration of a solution of potassium hydroxide (M = 56.0) containing 3.50 g KOH in 125 cm^3 of solution.

Step 1: Convert grams to moles: $\dfrac{3.50}{56.0} = 0.0625\,\text{mol KOH}$

Step 2: Change cm^3 to dm^3: $125\,cm^3 = 0.125\,dm^3$

Step 3: Calculate concentration: $\dfrac{0.0625}{0.125} = 0.500\,\text{mol dm}^{-3}\,\text{KOH}$

Worked example 2

Calculate the mass of magnesium chloride, $MgCl_2$ (M = 95.3) in 50 cm^3 of a 0.20 $mol\,dm^{-3}$ solution of magnesium chloride.

Step 1: Change cm^3 to dm^3: $50\,cm^3 = 0.050\,dm^3$

Step 2: Calculate number of moles, which equals: concentration ($mol\,dm^{-3}$) × volume (dm^3) = 0.20 × 0.050 = 0.010 mol $MgCl_2$

Step 3: Convert moles to grams: = 0.010 × 95.3 = 0.95 g (to two significant figures)

The answer is to two significant figures because the least number of significant figures in the data is two.

Worked example 3

What volume of a solution of concentration 0.10 mol dm^{-3} contains 2.0 g of sodium hydroxide, NaOH (M = 40)?

Step 1: Convert grams to moles: $\frac{2.0}{40} = 5 \times 10^{-2}$ mol NaOH

Step 2: Calculate volume in dm^3: $= \frac{\text{moles (mol)}}{\text{concentration (mol dm}^{-3})}$

$$= \frac{5 \times 10^{-2}}{0.10}$$

$$= 0.50\,\text{dm}^3$$

Acid–base titrations

A titration is used to determine the amount of substance present in a solution of unknown concentration. The procedure for determining the concentration of a solution of alkali is:

- Fill a burette with acid of known concentration (after washing it with the acid).
- Record the initial burette reading.
- Put a known volume of alkali into the flask using a volumetric pipette.
- Add an acid–base indicator to the alkali in the flask.
- Add the acid slowly from the burette until the indicator changes colour (end point).
- Record the final burette reading (final – initial burette reading is called the titre).
- Repeat the process until two or three successive titres differ by no more than 0.10 cm^3.
- Take the average of these titres. We usually ignore the first titre, since this is the rangefinder titration (rough titration). For example, in the table below, the fourth and fifth titres would be selected and averaged.

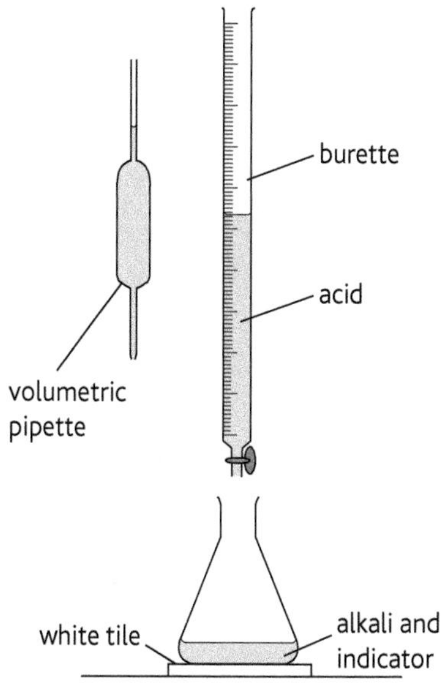

Figure 3.5.1 *The apparatus used in an acid–alkali titration*

First titre (rangefinder) (cm³)	Second titre (cm³)	Third titre (cm³)	Fourth titre (cm³)	5th titre (cm³)
32.92	32.85	32.00	32.25	32.35

Key points

- Molar concentration (mol dm^{-3}) = number of moles of solute (mol) ÷ volume of solution (dm^3).
- Acid–base titrations are carried out using an indicator which changes colour rapidly at the end point.
- When processing titration results, the values selected should be from two or three successive titres whose values are no more than 0.10 cm^3 apart.
- The answer to a calculation is expressed to the same number of significant figures as the lowest number of significant figures in the data provided.

3.6 More about titrations

Learning outcomes

On completion of this section, you should be able to:

- know how to calculate solution concentration from the results of titration experiments
- know how to use the results from a titration to calculate the stoichiometry of a reaction
- know some examples of redox titrations.

Calculating solution concentration by titration

In order to calculate solution concentration from titration results we need to know:

- the volume and concentration of the solution in the burette, e.g. the acid
- the volume of the solution in the flask
- the balanced equation for the reaction.

Worked example 1

25.0 cm³ of a solution of sodium hydroxide is exactly neutralised by 12.2 cm³ of aqueous sulphuric acid of concentration 0.100 mol dm^{-3}. Calculate the concentration in mol dm^{-3} of the sodium hydroxide solution.

Step 1: Calculate the moles of acid (moles = concentration × volume in dm³)

$$0.100 \times \frac{12.2}{1000} = 1.22 \times 10^{-3}\,\text{mol } H_2SO_4$$

Step 2: Use the stoichiometry of the equation to calculate moles of NaOH:

$$\begin{array}{cccc} H_2SO_4(aq) & + \; 2NaOH(aq) & \rightarrow Na_2SO_4\,(aq) + 2H_2O(l) \\ 1\text{ mol} & 2\text{ mol} & \\ 1.22 \times 10^{-3}\text{ mol} & 2.44 \times 10^{-3}\text{ mol} & \end{array}$$

Step 3: Calculate the concentration of NaOH (moles ÷ volume in dm³):

$$= \frac{2.44 \times 10^{-3}}{0.0250} = 0.0976\,\text{mol dm}^{-3}$$

Using titration data to deduce stoichiometry

To find the stoichiometry we need to know the concentrations and volumes of both reactants.

Worked example 2

25 cm³ of a 0.0250 mol dm^{-3} solution of a metal hydroxide was titrated with 0.10 mol dm^{-3} hydrochloric acid. It required 12.5 cm³ of acid to neutralise the hydroxide. Deduce the stoichiometry of this reaction.

Step 1: Calculate the number of moles of each reactant:

$$\text{moles hydroxide} = 0.0250 \times \frac{25}{1000} = 6.25 \times 10^{-4}\,\text{mol}$$

$$\text{moles of hydrochloric acid} = 0.10 \times \frac{12.5}{1000} = 1.25 \times 10^{-3}\,\text{mol}$$

Step 2: Deduce the simplest mole ratio of hydroxide to hydrochloric acid:

$$\begin{array}{ccccc} \text{hydroxide } 6.25 \times 10^{-4}\text{ mol} & : & \text{hydrochloric acid } 1.25 \times 10^{-3}\text{ mol} \\ OH^- \quad 1 & : & HCl \quad 2 \end{array}$$

Step 3: Write the stoichiometric equation:
$M(OH)_2 + 2HCl \rightarrow MCl_2 + 2H_2O$

Redox titrations

Redox titrations involve **oxidation–reduction** reactions (see Section 4.1). The end point in these titrations is detected either using the colour change of one of the reactants or by use of a special redox indicator. Some examples are given below:

Potassium manganate(VII) titrations

This can be used to find the concentration of, for example, Fe^{2+} ions in ammonium iron(II) sulphate. The reaction is carried out in the presence of acid (H^+ ions):

$$MnO_4^-(aq) + 5Fe^{2+}(aq) + 8H^+(aq) \rightarrow Mn^{2+}(aq) + 5Fe^{3+}(aq) + 4H_2O(l)$$

MnO_4^-: purple; Fe^{2+}: very light green; Mn^{2+}: very pale pink; Fe^{3+}: very pale yellow

The MnO_4^- ions in the potassium manganate(VII) are deep purple in colour. When the potassium manganate(VII) is added to the Fe^{2+} ions, the MnO_4^- ions are converted to Mn^{2+} ions which are almost colourless. After all the Fe^{2+} has reacted, the purple colour of the MnO_4^- ions are visible. So the colour change at the end point is yellowish (because of the Fe^{3+} ions) to purple.

Iodine–thiosulphate titrations

These are used in reactions where iodine is liberated, e.g. to find the concentration of a solution of hydrogen peroxide by adding iodide ions:

$$2I^-(aq) + H_2O_2(aq) + 2H^+(aq) \rightarrow I_2(aq) + 2H_2O(l)$$

The iodine liberated in this reaction as an aqueous solution, is titrated with a solution of sodium thiosulphate, $Na_2S_2O_3$:

$$I_2(aq) + 2Na_2S_2O_3(aq) \rightarrow 2NaI(aq) + Na_2S_4O_6(aq)$$

I_2: brown; $Na_2S_2O_3$: colourless; NaI: colourless; $Na_2S_4O_6$: colourless

The end point of the sodium thiosulphate titration occurs when the iodine present is used up. The colour change is from brown to colourless. The end point is sharpened by adding a few drops of starch solution to the titration flask as the end point is neared. The starch solution is blue-black in the presence of iodine and colourless when not. This blue-black to colourless change gives a sharper end point.

Dichromate(VI) titrations

The reactions of potassium dichromate are carried out in the presence of acid. For example:

$$Cr_2O_7^{2-}(aq) + 6Fe^{2+}(aq) + 14H^+(aq) \rightarrow 2Cr^{3+}(aq) + 6Fe^{3+}(aq) + 7H_2O(aq)$$

$Cr_2O_7^{2-}$: orange; Fe^{2+}: very light green; Cr^{3+}: green; Fe^{3+}: very pale yellow

The $Cr_2O_7^{2-}$ ions in the potassium dichromate(VI) are orange in colour. When the potassium dichromate is added to the Fe^{2+} ions, the $Cr_2O_7^{2-}$ ions are converted to Cr^{3+} ions which are green. The end point of the titration is not obvious so a redox indicator (barium diphenylamine sulphonate + phosphoric acid) is added. The colour change at the end point is then from green to violet-blue.

Did you know?

When using solutions of potassium manganate(VII) for titrations, you cannot keep the solution for very long because brown deposits of manganese(IV) oxide are formed. It is also not easy to see the meniscus in the burette because potassium manganate(VII) is one of the most intensely-coloured compounds known. For titrations involving this compound it is permissible to take burette readings from the top of the meniscus.

Key points

- The concentration of a solution can be found by titration using the relationship concentration ($mol\,dm^{-3}$) = number of moles of solute (mol) ÷ volume of solution (dm^3) together with the stoichiometric equation from the reaction.
- The stoichiometry of a reaction can be found by titration if the concentrations of both solutions are known.
- Redox titrations often involve a change in colour of one of the reactants.

Revision questions

1 Balance the following molecular equations, then write the ionic equation for each:

a $FeCl_3(aq) + NaOH(aq) \rightarrow Fe(OH)_3(s) + NaCl(aq)$

b $Mg(s) + HCl(aq) \rightarrow MgCl_2(aq) + H_2(g)$

c $Na_2S_2O_3(aq) + HCl(aq) \rightarrow NaCl(aq) + H_2O(l) + S(s) + SO_2(g)$

2 a Define:

i a mole

ii molar mass

b Calculate the molar mass of the following compounds:

i $Pb(NO_3)_2$

ii $Ca_3(PO_4)_2$

iii $CuSO_4{\cdot}5H_2O$

3 a Calculate the number of moles in the following:

i 7.1 g Na_2SO_4

ii 20 g $CaCO_3$

b Calculate the mass of the following:

i 0.025 mol $NaHCO_3$

ii 2.0×10^{-3} mol HCl

4 Calculate the mass of oxygen required to completely burn 5.5 g of propane (C_3H_8).

5 Write the molecular and empirical formula of the following compounds:

a dinitrogen tetroxide

b disulphur dichloride

6 A compound contains 0.48 g of carbon and 0.08 g of hydrogen and has a molar mass of 56 g mol^{-1}. Find the empirical and molecular formulae of the compound.

7 Calculate the masses of the following volumes of gases at r.t.p.

a 6 dm^3 of O_2

b 120 cm^3 of SO_2

8 When butane (C_4H_{10}) is burnt completely in oxygen, steam and carbon dioxide are produced. Calculate the volume of steam produced at s.t.p. when 56 cm^3 of butane is completely burnt.

9 Pentane contains carbon and hydrogen only. When 20 cm^3 of pentane reacts with 160 cm^3 of oxygen, 100 cm^3 of carbon dioxide is produced, the other product is steam.

Deduce the molecular formula of pentane.

10 Calculate the molar concentrations of the following solutions:

a 3.33 g KNO_3 in 250 cm^3 of solution

b 2.65 g Na_2CO_3 in 75 cm^3 of solution

11 Calculate the mass of solute in the following solutions:

a 25 cm^3 of a 0.01 mol dm^{-3} solution of NaCl

b 750 cm^3 of a 0.27 mol dm^{-3} solution of NH_4NO_3

12 A student titrated a standard solution of hydrochloric acid against 25 cm^3 of sodium hydroxide solution and obtained the following volumes of acid:

Burette readings/cm^3	**1**	**2**	**3**	**4**
Final volume		25.80	37.90	
Initial volume	0.50		12.50	1.00
Actual volume titre	26.65	25.65		25.50

a Complete the table above.

b From the results in the table, determine the average volume of acid required to neutralise the sodium hydroxide.

13 A student titrated a 0.2 mol dm^{-3} solution of potassium manganate(VII) against 25.0 cm^3 of iron(II) solution. The average volume of potassium manganate(VII) used in this titration was 27.50 cm^3.

Calculate the molar concentration of the iron(II) solution.

4 Redox reactions

4.1 Redox reactions and oxidation state

Learning outcomes

On completion of this section, you should be able to:

- understand redox reactions in terms of electron transfer and change in oxidation state (oxidation number)
- know the oxidation number rules
- deduce an oxidation number of an atom in a compound using oxidation number rules.

Exam tips

The phrase OIL RIG will help you to remember redox reactions in terms of electron transfer:

Oxidation Is Loss (of electrons).

Reduction Is Gain (of electrons).

What are redox reactions?

A simple definition of oxidation is the gain of oxygen. A simple definition of reduction is the loss of oxygen. But if we look at the reaction of copper(II) oxide with hydrogen, we can see that oxidation and reduction are taking place at the same time:

$$CuO(aq) + H_2(g) \rightarrow Cu(s) + H_2O(l)$$

Copper(II) oxide is losing oxygen and gets reduced. Hydrogen gains oxygen and gets oxidised. **Redox reactions** are reactions where oxidation and reduction are both taking place at the same time.

There are two other ways of finding out whether or not a substance has been oxidised or reduced:

- electron transfer
- changes in oxidation state.

Electron transfer in redox reactions

- Oxidation is loss of electrons.
- Reduction is gain of electrons.

Magnesium reacts with chlorine to form magnesium chloride:

$$Mg + Cl_2 \rightarrow MgCl_2$$

Each magnesium atom loses two electrons from its outer shell. This is oxidation.

$$Mg \rightarrow Mg^{2+} + 2e^-$$

Each chlorine atom in the chlorine molecule gains one electron to complete its outer shell. This is reduction.

$$Cl_2 + 2e^- \rightarrow 2Cl^-$$

Equations like these showing the oxidation or reduction reactions separately are called **half equations**. It is also acceptable, though not as preferable, to write these half equations as:

$$Mg - 2e^- \rightarrow Mg^{2+} \quad \text{and} \quad Cl_2 \rightarrow 2Cl^- - 2e^-$$

Oxidation states (oxidation numbers)

We can extend the definition of redox to include oxidation and reduction reactions involving covalent compounds by using oxidation states (oxidation numbers).

An **oxidation state** (**oxidation number**) is a number given to each atom or ion in a compound to show the degree of oxidation. In the following discussion oxidation number has been abbreviated as 'OxNo'.

Oxidation number rules

1 OxNo refers to a single atom or ion in a compound.

2 The OxNo of each atom in an element is 0, e.g. Mg = 0, H = 0.

3 The OxNo of an ion arising from a single atom is the same as the charge on the ion, e.g. $Mg^{2+} = +2$, $Fe^{3+} = +3$, $Br^- = -1$, $S^{2-} = -2$.

4 The OxNo of fluorine in compounds is always –1.

5 The OxNo of an oxygen atom in a compound is –2 (but in peroxides it is –1).

6 The OxNo of a hydrogen atom in a compound is +1 (but in metal hydrides it is –1), e.g. CH_4: H = +1, HCl: H = +1, NaH: H = –1.

7 The sum of all the OxNo's of atoms/ions in a compound is zero, e.g. in Al_2O_3

$$2Al^{3+} = 2 \times (+3) = +6 \qquad 3O^{2-} = 3 \times (-2) = -6$$

8 The sum of the OxNos in a compound ion equals the charge on the ion, e.g. in the NO^{3-} ion, OxNo of N = +5 and OxNo of 3 O atoms = 3 × (–2) = –6 +5 – 6 = –1 (the charge on the ion)

9 The most electronegative element in a compound is given the negative OxNo.

Applying oxidation number rules

The atoms of many elements in Groups V to VII have variable oxidation numbers. We work these out using the oxidation number rules. For example, what is the oxidation number of S in H_2SO_4?

- Applying rules 5 and 6: H = +1 and O = –2
- Applying rule 7: 2H + 4O + S = 0 so 2(+1) + 4(–2) + S = 0
- So OxNo of S = +6

Redox reactions and oxidation state

- Increase in oxidation number of an atom or ion in a reaction is oxidation.
- Decrease in oxidation number of an atom or ion in a reaction is reduction.

When copper(II) oxide reacts with ammonia, each atom of copper and nitrogen change as shown.

N oxidised

$$\underset{+2}{3CuO} + \underset{-3}{2NH_3} \longrightarrow \underset{0}{3Cu} + \underset{0}{N_2} + 3H_2O$$

Cu reduced

The copper in the copper oxide is reduced to copper (OxNo change from +2 to 0).

The nitrogen in the ammonia is oxidised to nitrogen gas (OxNo change from –3 to 0).

Did you know?

The Roman numbers after the names of particular atoms (or ions) in compounds show the oxidation state of that particular atom (or ion). This is called the 'Stock nomenclature'. Iron(II) chloride shows that iron has an OxNo of +2. Potassium manganate(VII) shows that the manganese has an OxNo of +7. Chlorine(I) oxide shows that chlorine has an OxNo of +1.

Key points

- Oxidation is loss of electrons. Reduction is gain of electrons.
- Redox equations can be divided into two half equations, one showing oxidation and the other reduction.
- Oxidation is increase in oxidation state. Reduction is decrease in oxidation state.
- Some atoms have fixed oxidation numbers. The oxidation number of other atoms in a compound can be found using oxidation number rules.

4.2 Redox equations

Learning outcomes

On completion of this section, you should be able to:

- deduce balanced equations from relevant half equations
- understand the terms oxidising agent and reducing agent
- know how to order elements in terms of oxidising/ reducing ability by reference to displacement reactions.

Balancing half equations

We balance two half equations by:

- balancing the number of electrons lost and gained
- then putting the two half equations together.

Worked example 1

Construct a balanced ionic equation for the reaction of zinc with iodate(V) ions, IO_3^-, in acidic solution using the two half equations shown below:

Equation 1: Oxidation of zinc to zinc(II) ions: $Zn \rightarrow Zn^{2+} + 2e^-$

Equation 2: Reduction of iodate(V) to iodine:

$$IO_3^- + 6H^+ + 5e^- \rightarrow \tfrac{1}{2}I_2 + 3H_2O$$

- Each Zn atom loses 2 electrons when oxidised. Each I atom in the IO_3^- ion gains 5 electrons when reduced.
- To balance the electrons:
  Multiply equation 1 by 5 (10 electrons in total):

$$5Zn \rightarrow 5Zn^{2+} + 10e^-$$

  Multiply equation 2 by 2 (10 electrons in total):

$$2IO_3^- + 12H^+ + 10e^- \rightarrow I_2 + 6H_2O$$

- Put the two equations together and cancel the electrons:

$$5Zn + 2IO_3^- + 12H^+ \rightarrow 5Zn^{2+} + I_2 + 6H_2O$$

Balancing equations using oxidation numbers

Worked example 2

Write a balanced equation for the reaction of ClO_3^- ions in acidic solution with Fe^{2+} ions to form Cl^- ions, Fe^{3+} ions and water.

Step 1: Write down the unbalanced equation, identify the atoms which change in oxidation number.

$$ClO_3^- + Fe^{2+} + H^+ \rightarrow Cl^- + Fe^{3+} + H_2O$$

Step 2: Deduce the changes in oxidation number.

OxNo change −6

$$\underset{+5}{ClO_3^-} + \underset{+2}{Fe^{2+}} + H^+ \longrightarrow \underset{-1}{Cl^-} + \underset{+3}{Fe^{3+}} + H_2O$$

OxNo change +1

Step 3: Balance the oxidation number changes so that the total oxidation number is 0.

$$ClO_3^- + 6Fe^{2+} + H^+ \rightarrow Cl^- + 6Fe^{3+} + H_2O$$

Step 4: Balance the charges by balancing the H^+ ions (or OH^- ions if present) and then other molecules such as water.

$$ClO_3^- + 6Fe^{2+} + 6H^+ \rightarrow Cl^- + 6Fe^{3+} + 3H_2O$$

Exam tips

- When balancing oxidation numbers, the oxidation numbers are per atom – you must compare the same number of atoms on each side of the equation.
- If a question states that acid is present, write H^+ on the left.
- When balancing using oxidation numbers, remember you usually need to select two reactants and two products whose oxidation numbers change. If the reaction is a **disproportionation reaction** (see Section 12.8) you will only need to select three species.

Reducing agents and oxidising agents

During a redox reaction:

- A reducing agent loses electrons and gets oxidised.
- An oxidising agent gains electrons and gets reduced.

The ability of a substance to act as a reducing agent or oxidising agent depends on its electrode potential (see Section 10.1). We can put reagents in order of their oxidising or reducing ability by carrying out displacement reactions. A **displacement reaction** is one where one atom replaces another in a chemical reaction. For example, zinc displaces copper from an aqueous solution of copper(II) sulphate:

$$Zn(s) + Cu^{2+}(aq) \rightarrow Zn^{2+}(aq) + Cu(s)$$

Zinc is a better reducing agent than copper (Zn is better at losing electrons). Cu^{2+} is a better oxidising agent than Zn^{2+} (Cu^{2+} is better at gaining electrons). So the reaction goes to the right.

Silver does not displace copper from copper(II) sulphate solution because silver is a better oxidising agent than copper and the Cu^{2+} ion is a better reducing agent than Ag^{+}.

If we react different metals with aqueous solutions of different metal salts, we can build up a metal **reactivity series**, e.g.

⟵ metal is a better **reductant** (better at releasing electrons) ⟵

Mg Zn Fe Sn Cu Ag

⟶ metal ion is a better **oxidant** (better at accepting electrons) ⟶

According to this reactivity series, magnesium will displace silver from silver nitrate.

$$Mg(s) + 2AgNO_3(aq) \rightarrow Mg(NO_3)_2(aq) + 2Ag(s)$$

But silver will not displace magnesium from aqueous magnesium nitrate.

Displacement reactions can also involve non-metals. A more reactive halogen will displace a less reactive one from an aqueous solution of halide ions (see Section 12.6).

Key points

- Redox equations can be deduced by balancing the numbers of electrons lost and gained in relevant half equations.
- Redox equations can be balanced by balancing oxidation number changes.
- A more reactive element will displace a less reactive element from a solution of its salt if it is better at losing electrons.
- In a reaction, the oxidising agent gains electrons and becomes reduced. The reducing agent loses electrons and becomes oxidised.

5 Kinetic theory

5.1 The gas laws

Learning outcomes

On completion of this section, you should be able to:

- state Boyle's law and Charles' law
- draw graphical representations to illustrate Boyle's law and Charles' law
- calculate volumes and pressures of gases under different conditions using Boyle's law
- calculate volumes and temperatures of gases under different conditions using Charles' law.

Did you know?

The chemical industry often uses very high pressures in industrial processes involving gases, e.g. in the Haber process for making ammonia. In such cases it is convenient to use 'atmospheres' (abbreviated atm) as the unit of pressure. This is because it is easier to deal with smaller numbers than larger: 1 atm = 101 000 Pa.

Compressing gases

We can picture gases as a collection of randomly moving particles which are continuously colliding with each other and with the walls of the container in which they are placed. The gas particles exert a pressure because they are constantly hitting the walls of the container:

$$\text{pressure } (\text{N m}^{-2}) = \frac{\text{force (N)}}{\text{area } (\text{m}^2)}$$

When we decrease the volume of a gas, the molecules get squashed closer together. The molecules hit the walls of the container more often and so the gas pressure increases.

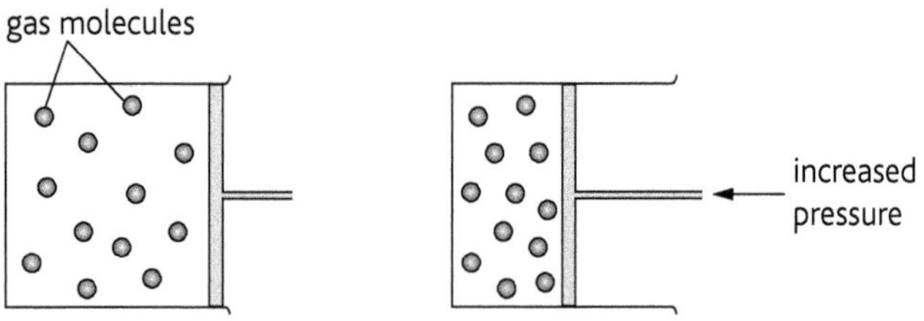

Figure 5.1.1 *When the volume of the container is decreased, the gas molecules are squashed closer together and hit the walls of the container more often*

We measure pressure in newtons per square metre (N m^{-2}) or in pascals

$$1\,\text{N m}^{-2} = 1 \text{ pascal (Pa)}.$$

Standard atmospheric pressure is 101 325 Pa (often rounded down to 101 000 Pa).

Boyle's law

Boyle's law states: *For a fixed number of moles of a gas at a fixed temperature, the volume of a gas is inversely proportional to its pressure.* Boyle's law is obeyed as long as the temperature is not too low or the pressure too high. Figure 5.1.2 shows two graphical representations of Boyle's law.

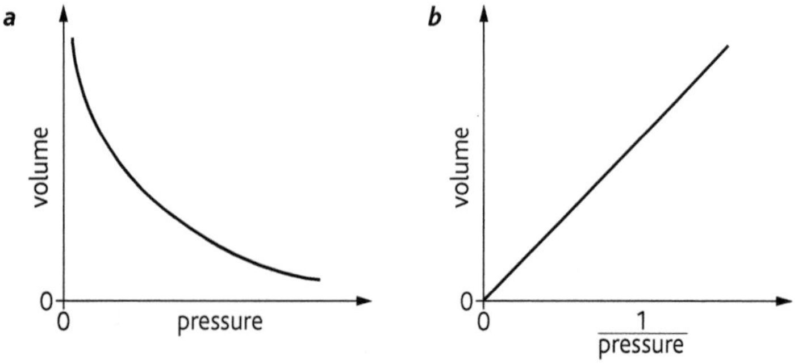

Figure 5.1.2 *Boyle's law shown graphically:* ***a*** *As the volume of gas decreases the pressure increases;* ***b*** *A plot of volume against 1/pressure shows proportionality*

Under different conditions we can represent Boyle's law mathematically as:

$$\underset{\text{initial pressure} \times \text{initial volume}}{P_1V_1} = \underset{\text{final pressure} \times \text{final volume}}{P_2V_2}$$

Worked example 1

A gas has a volume of 1.2 dm³. Its pressure is 100 kPa. What is the volume of this gas when the pressure is increased to 250 kPa at constant temperature?

Step 1: Substitute the values into $P_1V_1 = P_2V_2$: $100 \times 1.2 = 250 \times V_2$

Step 2: Calculate the new volume: $V_2 = \frac{100 \times 1.2}{250} = 0.48\,\text{dm}^3$

Heating gases

As the temperature of a gas increases (at a fixed pressure), the molecules move faster and hit the wall with increased force because they have more energy. If the pressure is to remain constant, the molecules must get further apart. So the volume of gas increases as the temperature increases. If we decrease the temperature of a gas, we must eventually reach a point when the theoretical volume of the gas is zero. This temperature is called the absolute temperature.

In chemistry we often use the absolute temperature scale (Kelvin temperature scale). The units of this scale are kelvins (K).

Kelvin temperature = °C + 273.
So minus 15 °C in kelvin is –15 + 273 = 258 K.

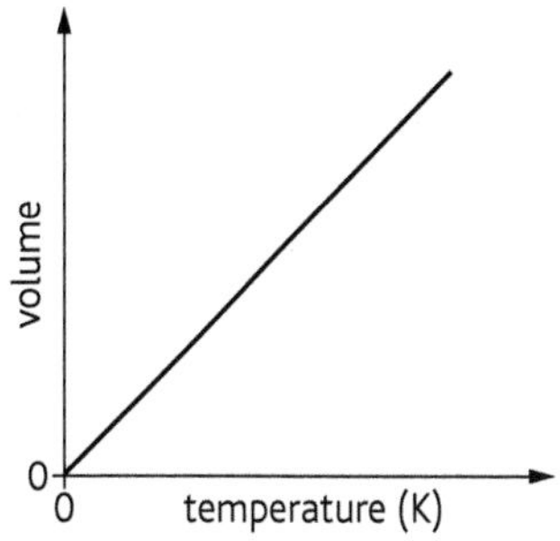

Figure 5.1.3 *As the temperature of a gas increases, its volume increases in direct proportion (Charles' law)*

Charles' law

Charles' law states: *For a fixed number of moles of a gas at constant pressure, the volume of a gas is directly proportional to its absolute (kelvin) temperature.* Figure 5.1.3 shows a graphical representation of Charles' law.

Under different conditions we can represent Charles' law mathematically as:

$$\frac{\text{initial volume}}{\text{initial temperature (K)}} = \frac{\text{final volume}}{\text{final temperature (K)}} \quad \frac{V_1}{T_1} = \frac{V_2}{T_2}$$

Worked example 2

At 27 °C a gas occupies a volume of 600 cm³. What volume does the gas occupy at 227 °C assuming that the pressure remains constant?

Step 1: Convert °C to K:

$$27\,°\text{C} = 27 + 273 = 300\,\text{K}$$
$$227\,°\text{C} = 227 + 273 = 500\,\text{K}$$

Step 2: Substitute the values into the equation $\frac{V_1}{T_1} = \frac{V_2}{T_2}$: $\frac{600}{300} = \frac{V_2}{500}$

Step 3: Calculate the new volume: $V_2 = \frac{600}{300} \times 500 = 1000\,\text{cm}^3$

Exam tips

You can combine Boyle's law and Charles' law into one equation if pressure and temperature both change:

$$\frac{P_1V_1}{T_1} = \frac{P_2V_2}{T_2}$$

Key points

- Standard pressure and temperature are 101 325 Pa and 273 K respectively.
- Kelvin temperature = °C + 273
- Boyle's law: At constant temperature, the volume of a gas is inversely proportional to its pressure.
- Charles' law: At constant pressure, the volume of a gas is directly proportional to its absolute temperature.

5.2 The kinetic theory of gases

Learning outcomes

On completion of this section, you should be able to:

- know the characteristics of the gaseous state
- know the basic assumptions of the kinetic theory
- know the differences between real and ideal gases
- know the ideal gas equation.

General properties of gases

Most gases are either small molecules, e.g. ammonia and sulphur dioxide, or exist as isolated atoms, e.g. helium and argon.

The particles in a gas:

- are arranged randomly (are disordered)
- are far apart from each other
- move rapidly and randomly.

The kinetic theory and ideal gases

The **kinetic theory** of gases states that:

- Gas particles are moving randomly.
- Gas particles do not attract each other.
- Gas particles have no volume.
- Collisions between gas particles are elastic, i.e. no energy is lost when they collide.

A gas which has these characteristics is called an **ideal gas**. In such a gas, the kinetic energy of the particles is a measure of their temperature. Ideal gases obey Boyle's law and Charles' law exactly.

Real gases

Scientists have made accurate measurements of the volumes of many gases at different pressures and temperatures. The results show that many gases do not obey Boyle's law and Charles' law exactly. These differences are especially noticeable at very high pressures and very low temperatures. The kinetic theory is not always obeyed because:

- We cannot ignore the volumes of the particles.
- The attraction between the particles is not zero.

At very high pressure and very low temperature:

- The particles are closer together.
- So the volume of the particles is not negligible compared with the volume of the container they are placed in.
- The forces of attraction between the particles cannot be ignored.
- Attractive forces between the particles pull them towards each other.
- The pressure is lower than expected for an ideal gas and the effective volume of the gas is lower than expected.
- At extremely high pressures and extremely low temperatures the particles may be so close to one another that repulsive forces between the electrons clouds cause deviations from ideal gas behaviour.

Gases which do not obey Boyle's law and Charles' law at all temperatures and pressures are called **real gases**.

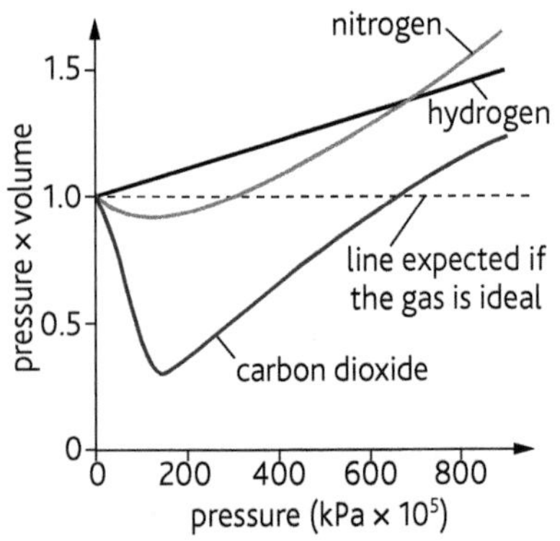

***Figure 5.2.1** Hydrogen, nitrogen and carbon dioxide show deviations from Boyle's law over a range of pressures*

The ideal gas equation

For an ideal gas we can combine Boyle's law and Charles' law. This gives us the **ideal gas equation**:

$$PV = nRT$$

Where:

P is the pressure in pascals, Pa.
V is the volume in cubic metres ($1\,m^3 = 1000\,dm^3$).
n is the number of moles of gas.
R is the **gas constant** ($8.31\,J\,K^{-1}\,mol^{-1}$).
T is the temperature in kelvin (K).

Since number of moles $= \dfrac{\text{mass in grams (m)}}{\text{molar mass (M)}}$ we can rewrite the ideal gas equation as:

$$PV = \frac{mRT}{M}$$

If we know four of the five quantities in the ideal gas equation, $PV = nRT$, we can calculate the fifth.

Key points

- The kinetic theory states that gas particles are always in constant random motion at a variety of speeds.
- An ideal gas is one which obeys the gas laws. Real gases do not obey the gas laws under all pressures and temperatures.
- Gases do not obey the gas laws at high pressures and low temperatures.
- The ideal gas equation is $PV = nRT$.

5.3 Using the ideal gas equation

Learning outcomes

On completion of this section, you should be able to:

- perform calculations using the ideal gas equation
- know how to determine the relative molecular mass of liquids with low boiling points or gases by inserting relevant data into the ideal gas equation.

Exam tips

1. When doing gas law calculations, always think of the units that you are likely to have to change to as 'molPaK'
   moles mol
   pressure Pa
   temperature K
2. Make sure that you change cm^3 or dm^3 to m^3 ($1\,m^3 = 1000\,dm^3$)

Calculations using the ideal gas equation

Worked example 1

Calculate the volume of gas in a weather balloon which contains 0.24 kg of helium at a temperature of –25 °C and a pressure of 60 kPa (A_r He = 4.0; $R = 8.31\,J\,K^{-1}\,mol^{-1}$)

Step 1: Change pressure and temperature into the correct units and calculate the number of moles:

$$60\,kPa = 60\,000\,Pa \qquad -25\,°C = -25 + 273 = 248\,K$$

$$0.24\,kg = 240\,g \qquad \text{moles of He} = \frac{240}{4} = 60\,mol$$

Step 2: Rearrange the ideal gas equation into the form you need:

$$PV = nRT \rightarrow V = \frac{nRT}{P}$$

Step 3: Substitute the figures: $V\ (\text{in } m^3) = \dfrac{60 \times 8.31 \times 248}{60\,000} = 2.06\,m^3$

Deducing relative molecular masses

We can use a simple weighing method to deduce the relative molecular mass of a gas.

A large flask of known volume is filled with gas and the mass of the gas is found. If the temperature and pressure are known, we can substitute all the values in the ideal gas equation to find the value of M. Although this method is suitable in the school laboratory when dense gases are being used, its accuracy is limited because of problems involving the buoyancy of air.

Worked example 2

A flask of volume $2.00\,dm^3$ contains 3.61 g of a gas. The pressure in the flask is 100 kPa and the temperature is 20 °C. Calculate the relative molecular mass of the gas ($R = 8.31\,J\,K^{-1}\,mol^{-1}$).

Step 1: Change pressure, volume and temperature to their correct units:

$$100\,kPa = 100\,000\,Pa \qquad 2.0\,dm^3 = 2.00/1000 = 2.00 \times 10^{-3}\,m^3$$

$$20\,°C = 20 + 273 = 293\,K$$

Step 2: Rearrange the ideal gas equation (in the form showing mass, m, and molar mass, M) to make M the subject:

$$PV = \frac{mRT}{M} \rightarrow M = \frac{mRT}{PV}$$

Step 3: Substitute the values:

$$M = \frac{3.61 \times 8.31 \times 293}{100\,000 \times 2.00 \times 10^{-3}} = 44 \text{ (to two significant figures)}$$

Deducing the relative molecular mass of a volatile liquid

A similar method can be used to deduce the molar mass of a low boiling point liquid.

The apparatus is shown below:

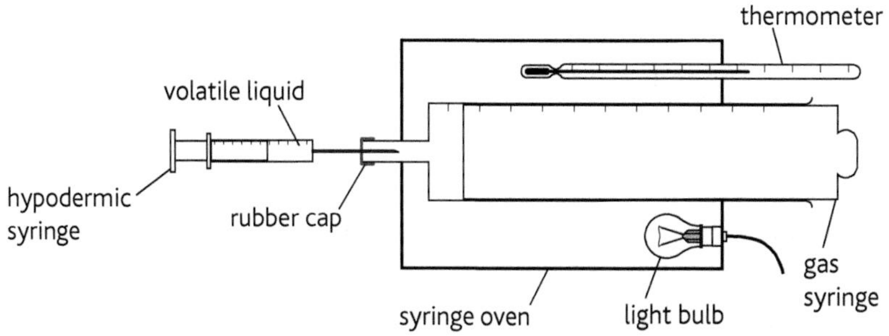

Figure 5.3.1 *Apparatus used for finding the relative molecular mass of a volatile liquid*

The procedure is:

- Let the gas syringe reach a particular temperature in the syringe oven. Record this temperature.
- Record volume of the air in the gas syringe.
- Draw some of the liquid into the hypodermic syringe and weigh it.
- Inject some of the liquid from the hypodermic syringe into the gas syringe.
- Allow the liquid to vaporise and record the final volume of air + vapour in the gas syringe.
- Record the mass of the hypodermic syringe and the pressure.

The calculation is carried out in the same way as for a gas but:

- Volume of vapour = final gas syringe volume – initial gas syringe volume
- Mass of liquid = initial mass of hypodermic syringe – final mass of hypodermic syringe

Key points

- The ideal gas equation, $PV = nRT$, can be used to deduce pressure, volume, temperature or number of moles if the other physical quantities in the equation are known.
- The relative molecular mass of a gas can be found by weighing a known volume of gas (at known temperature and pressure).
- The relative molecular mass of liquids with low boiling points can be found by measuring the volume of vapour produced when a known mass of liquid is vaporised (at known temperature and pressure).

5.4 Changing state

Learning outcomes

On completion of this section, you should be able to:

- describe the liquid state in terms of motion, proximity and arrangement of particles
- explain the terms 'melting' and 'vaporisation'
- explain changes in state in terms of energy changes and changes in proximity, arrangement and motion of particles.

Did you know?

Water in the liquid state has a great deal of structure. This is due to its extensive hydrogen bonding. Larger or smaller groups of water molecules exist in the structure, perhaps with up to 18 water molecules with an ice-like structure between more randomly arranged molecules. Hydrogen bonds are continuously being broken and formed within the structure. This is sometimes referred to as a flickering crystal structure of water.

The liquid state

The molecules in a liquid:

- are not regularly arranged, although there may be some order over short distances due to the effect of weak intermolecular forces of attraction which are constantly being broken up as they gain kinetic energy from neighbouring molecules
- are close together
- have more kinetic energy than in a solid and so are able to slide over each other in a fairly random way.

Most substances that are liquid at room temperature are either:

- covalently bonded molecules with considerable van der Waals forces of attraction, e.g. bromine
- molecules with significant hydrogen bonding, e.g. water or dipole–dipole bonding e.g. propanone.

Changing state

When a solid is heated:

- The energy transferred increases the vibration of the particles. The temperature of the solid increases.
- The forces of attraction between the particles weaken.
- The solid melts when enough energy is transferred to make the particles slide over each other.
- The melting point is the temperature when the solid is in equilibrium with the liquid.

For an ionic solid a lot of energy is required to break the strong ionic attractive forces in the lattice. So the melting point is very high. For a molecular solid, less energy is required for melting – just enough to overcome the weak intermolecular forces holding the molecules together.

When a liquid is heated:

- The energy transferred increases the movement and energy of the particles. The temperature of the liquid increases.
- The forces of attraction between the particles weaken until they escape from the liquid and become a vapour. This process is called **vaporisation**.
- The liquid boils when enough energy is transferred to allow an equilibrium between the liquid and the gas phase.

Exam tips

It is important to distinguish between evaporation and boiling. Even below the boiling point some particles have enough energy to escape from the liquid. This is called **evaporation**. Boiling occurs when the whole liquid is in equilibrium with the vapour, which appears as bubbles within the liquid.

These changes of state can be shown on a heating curve. The temperature stays constant at the melting point and at the boiling point because the energy absorbed goes to break the forces between the molecules (or ions) rather than raising the temperature.

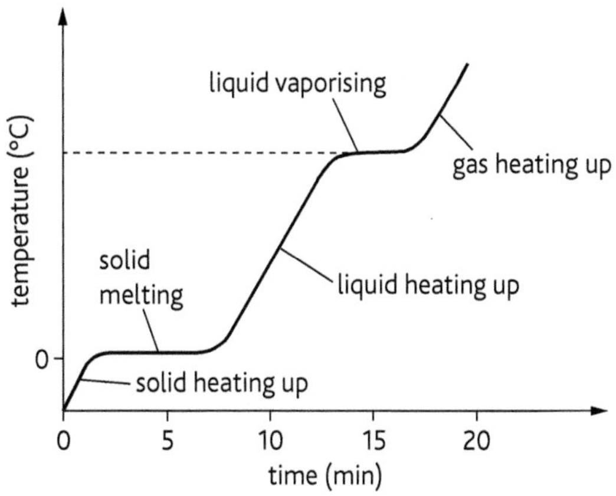

Figure 5.4.1 *The change in temperature when water is heated at a constant rate. Note how the temperature stays constant as the solid melts and the liquid turns to vapour*

When a vapour is cooled, the particles lose kinetic energy and begin to be attracted to each other. Energy is released and the vapour turns to liquid. It **condenses**.

When a liquid is cooled, the particles lose kinetic energy and slide past each other more and more slowly. Eventually the temperature is sufficiently low for the liquid to solidify. The liquid freezes.

Summarising changes of state

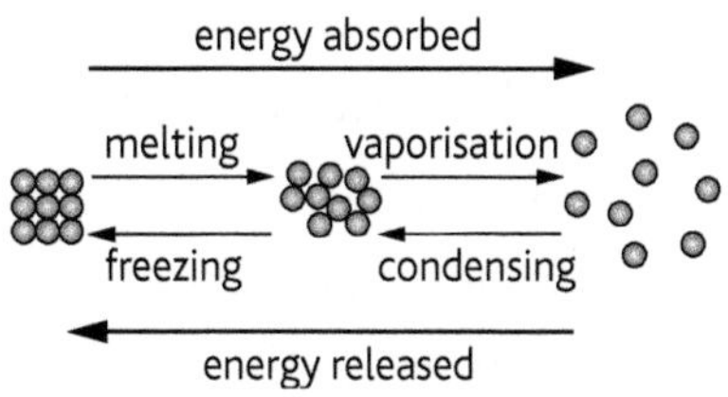

Figure 5.4.2 *Energy is required to melt and vaporise a substance. Energy is released when a substance condenses and freezes*

Key points

- Particles in a liquid are not ordered, are close together and move by sliding over each other.
- Most substances which are liquid at room temperature have a simple molecular structure with weak attractive forces between the molecules.
- Particles in a solid are generally arranged in a lattice. The particles only vibrate.
- When a solid melts, the energy input overcomes the forces keeping the particles together. The particles begin to slide over one another.
- When a liquid vaporises, the particles gain energy and move further apart. They move freely.

6 Energetics

6.1 Enthalpy changes

Learning outcomes

On completion of this section, you should be able to:

- know that energy changes accompany chemical reactions
- understand the meaning of the term 'enthalpy change'
- describe the terms 'exothermic' and 'endothermic'
- draw energy profile diagrams for exothermic and endothermic reactions.

Exam tips

In Chemistry it is useful to know some Greek letters and what they mean. You have already come across two: sigma and pi. The Greek letter capital delta Δ is used to denote a change in something. So ΔH means enthalpy change and ΔT means change in temperature.

Enthalpy and enthalpy change

Particles have two main types of energy:

- Potential energy is the energy due to the position of the particles. The closer two attracting particles are to each other, the lower the potential energy.
- Kinetic energy is the energy associated with the movement of particles.

Energy has to be absorbed to separate particles from each other. The heat content is the total amount of energy (potential and kinetic) in chemicals. The heat content is also called the **enthalpy** (symbol H).

We cannot measure enthalpy by itself. So it was decided that all elements in their normal physical states at 101.325 kPa pressure and 298 K have zero enthalpy. These conditions are called **standard conditions**. The symbol $^\ominus$ is used to show standard conditions.

We can, however, measure **enthalpy changes**. These occur when heat energy is exchanged with the surroundings in a chemical reaction. The symbol for standard enthalpy change is $\Delta H^\ominus$. The enthalpy change is the difference between the enthalpy of the products and the reactants.

$$\Delta H^\ominus = H^\ominus_{\text{products}} - H^\ominus_{\text{reactants}}$$

The units of enthalpy change are kJ mol^{-1}.

Exothermic and endothermic reactions

Two important terms which are common in chemistry are:

- **system** – the chemical reactants and products
- **surroundings** – these include the air, the container in which a reaction is carried out, any solvents not taking part in the reaction and anything dipping into the reaction mixture, e.g. a thermometer.

An **endothermic** reaction absorbs energy from the surroundings. The surroundings decrease in temperature.

$$H^\ominus_{\text{products}} - H^\ominus_{\text{reactants}} \text{ is positive. So } \Delta H^\ominus_{\text{reaction}} \text{ is positive.}$$

For example: $CaCO_3(s) \rightarrow CaO(s) + CO_2(g) \quad \Delta H^\ominus_{\text{reaction}} = +572\ \text{kJ mol}^{-1}$

All reactions requiring a continuous input of heat are endothermic, e.g. thermal decompositions. Some salts dissolve in water endothermically, e.g. ammonium chloride.

An **exothermic** reaction releases energy to the surroundings. The surroundings increase in temperature.

$$H^\ominus_{\text{products}} - H^\ominus_{\text{reactants}} \text{ is negative. So } \Delta H^\ominus_{\text{reaction}} \text{ is negative}$$

For example: $C(s) + O_2(g) \rightarrow CO_2(g)\ \Delta H^\ominus_{\text{reaction}} = -394\ \text{kJ mol}^{-1}$

All combustion reactions are exothermic. Many reactions of metals with acids are exothermic.

Energy profile diagrams

Energy profile diagrams (enthalpy profile diagrams) show:

- the relative enthalpy of the reactants and products on the *y*-axis, this usually includes the formulae of reactants and products
- the reaction pathway on the *x*-axis
- the enthalpy change, a downward arrow, ↓, indicates energy released (exothermic), an upward arrow, ↑, indicates energy absorbed (endothermic).

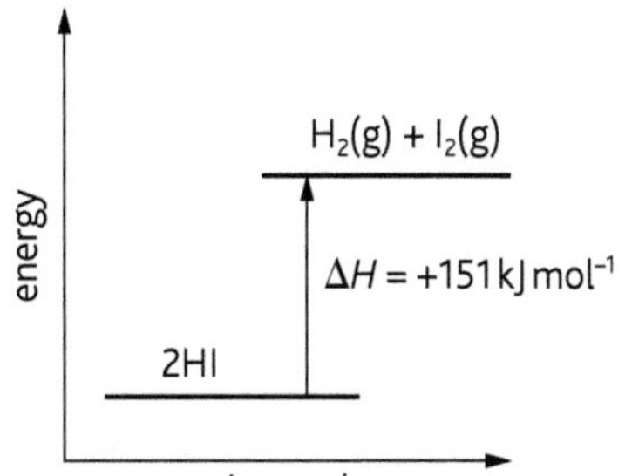

Figure 6.1.1 *Energy profile diagram for the thermal decomposition of hydrogen iodide (endothermic)*

The diagram for the endothermic reaction shows that:

- The reaction mixture gains energy so the surroundings cool down.
- So the enthalpy of the products is higher than that of the reactants.
- So $\Delta H^{\ominus}_{reaction}$ is positive.

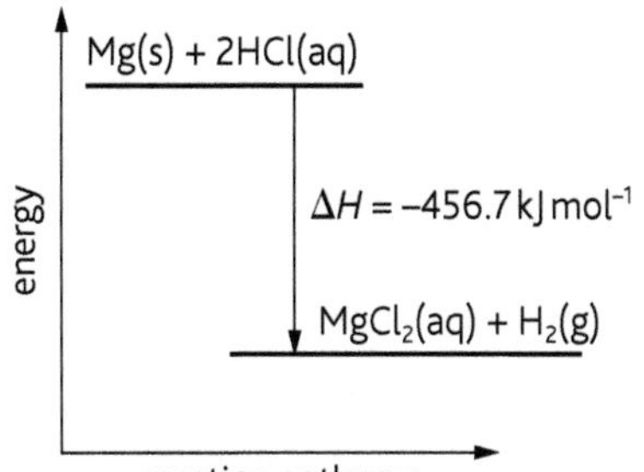

Figure 6.1.2 *Energy profile diagram for the reaction of magnesium with hydrochloric acid (exothermic)*

The diagram for the exothermic reaction shows that:

- The reaction mixture loses energy so the surroundings warm up.
- So the enthalpy of the reactants is higher than that of the products.
- So $\Delta H^{\ominus}_{reaction}$ is negative.

Key points

- Energy changes in chemical reactions result in heat loss or gain: an enthalpy change, ΔH.
- In an exothermic reaction, heat is transferred to the surroundings. In an endothermic reaction heat is absorbed from the surroundings.
- Exothermic reactions have negative ΔH values, endothermic reactions have positive ΔH values.
- Energy profile diagrams show the relative energies of the reactants and products and the enthalpy change.

6.2 Bond energies

Learning outcomes

On completion of this section, you should be able to:

- know that chemical reactions take place through energy changes involved in bond breaking and making
- know that bond breaking is endothermic and bond making is exothermic
- understand how bond energies can be used to calculate the enthalpy change of reaction
- describe the factors affecting bond energy and how bond energy can affect chemical reactivity.

Exam tips

Do not make the mistake in thinking that during a reaction all the bonds in the compound are broken into atoms. In many reactions only certain bonds break during the reaction. Particular bonds break and form in a particular sequence (see Section 7.5 for more details). Energy level diagrams showing all the atoms breaking then reforming are for 'atom accounting' only.

Bond making and bond breaking

- Breaking the bond between two atoms requires energy. It is endothermic.
- Making new bonds between two atoms releases energy. It is exothermic.
- If the energy absorbed in bond breaking is greater than the energy released in making new bonds, the reaction is endothermic.
- If the energy absorbed in bond breaking is less than the energy released in making new bonds, the reaction is exothermic.

We can draw energy profile diagrams to show these changes.

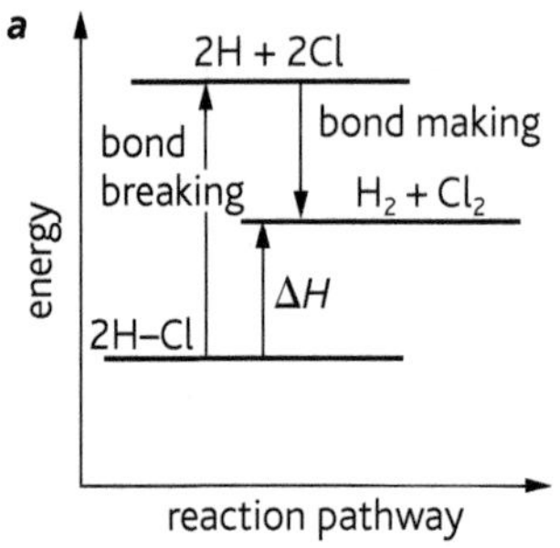

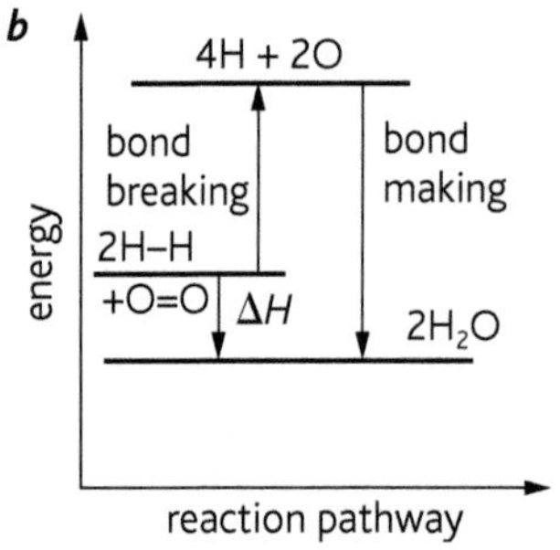

Figure 6.2.1 *Energy profile diagrams for:* ***a*** *an endothermic reaction – the thermal decomposition of hydrogen chloride;* ***b*** *an exothermic reaction – the synthesis of water from hydrogen and oxygen*

Bond energies

Bond energy is the amount of energy needed to break one mole of a particular bond in one mole of gaseous molecules. This is also called bond **dissociation** energy. The symbol for bond energy is E. The type of bond broken is put after this symbol.

The units of bond energy are kJ mol^{-1}. So $E(\text{C=C}) = +610\,\text{kJ mol}^{-1}$ refers to the energy required to break a C=C double bond.

Values of bond energies are always positive because they refer to bonds being broken. When new bonds are formed, the amount of energy released is the same as the amount of energy absorbed when the same type of bond is broken. For example:

$$\text{O=O(g)} \rightarrow 2\text{O(g)} \quad \Delta H = +\,496\,\text{kJ mol}^{-1}$$
$$2\text{O(g)} \rightarrow \text{O=O(g)} \quad \Delta H = -\,496\,\text{kJ mol}^{-1}$$

Factors affecting bond energy

For atoms of the same type, double bonds have higher bond energies than single bonds. This is because there is a greater attractive force between two bonding pairs of electrons and the nuclei of the atoms forming the bond compared with only one electron pair.

$$E(\text{C–C}) = +\,350\,\text{kJ mol}^{-1} \qquad E(\text{C=C}) = +\,610\,\text{kJ mol}^{-1}$$
$$E(\text{C≡C}) = +840\,\text{kJ mol}^{-1}$$

The bond energies of the halogens are:

$$E(\text{Cl–Cl}) = +\,244\,\text{kJ mol}^{-1} \qquad E\,(\text{Br–Br}) = +\,193\,\text{kJ mol}^{-1}$$
$$E(\text{I–I}) = +\,151\,\text{kJ mol}^{-1}$$

The bond energies decrease down the group as the distance between the atomic nuclei increases. This is because there is less force of attraction between the bonding pair of electrons and the nuclei as the bond lengthens. The bond energy for fluorine, however, is much lower than expected ($E(\text{F–F}) = +\,158\,\text{kJ mol}^{-1}$). This is because the F atoms are very close to each other so the electron clouds in the neighbouring atoms repel each other considerably.

The bond energies of carbon–halogen bonds also decrease as the halogen atom increases in size.

$$E(\text{C–Cl}) = +340\,\text{kJ mol}^{-1} \qquad E(\text{C–Br}) = +280\,\text{kJ mol}^{-1}$$
$$E(\text{C–I}) = +240\,\text{kJ mol}^{-1}$$

The bond energy of a particular type of bond can be affected by other atoms in a molecule. For example:

$$E(\text{C–F})\text{ in }CH_3F = +\,452\,\text{kJ mol}^{-1} \qquad E(\text{C–F})\text{ in }CF_4 = +\,485\,\text{kJ mol}^{-1}$$

For this reason we often use average bond energies, $\bar{E}$.

Calculating enthalpy changes using bond energies

The 'balance sheet method' for calculating enthalpy changes using bond energies is shown below for the reaction:

```
  H
  |
H–C–H + 2 O=O → O=C=O + 2 H–O–H
  |
  H
```

Bonds broken (endothermic +) /kJ mol^{-1}	Bonds formed (exothermic –) /kJ mol^{-1}
4 × E(C–H) = 4 × 410 = 1640 2 × E(O=O) = 2 × 496 = 992	2 × E(C=O) = 2 × 740 = –1480 4 × E(O–H) = 4 × 460 = –1840
Total = +2632 kJ	Total = –3320 kJ

Overall enthalpy change = +2632 – 3320 = –688 kJ mol^{-1}

Bond energy and reactivity

The bond energy of the N≡N triple bond is very high (994 kJ mol^{-1}). Nitrogen only reacts under very harsh conditions because it is difficult to break the bonds to get the reaction started. Hydrogen peroxide decomposes very readily because the O–O bond energy is only 150 kJ mol^{-1}. The ease of decomposition of hydrogen halides is also related to the strength of the hydrogen–halogen bond (see Section 12.7).

Did you know?

In compounds such as ethene the pi bond in the C=C bond is weaker than the sigma bond. You can see this if you compare the bond energies of the carbon–carbon single and double bonds. This reflects the more exposed electron density of the pi bond and the smaller amount of overlap of the p_z atomic orbitals compared with the sp^2 bonding in the sigma molecular orbitals. (see Section 2.9).

Key points

- Bond breaking is an endothermic process and bond making is exothermic.
- Bond energy is the energy needed to break one mole of a particular bond in one mole of gaseous molecules.
- Bond energies can be used to calculate the enthalpy change in a reaction.
- The strength of the bond between two given atoms is slightly different in different compounds.
- A high value of bond energy may contribute to a molecule being relatively unreactive.

6.3 Enthalpy changes by experiment

Learning outcomes

On completion of this section, you should be able to:

- describe enthalpy changes of reaction, combustion, neutralisation and solution
- know how to determine these enthalpy changes by experiment
- use experimental data to calculate enthalpy changes.

Four types of enthalpy change

- **Standard enthalpy change of reaction:** The enthalpy change when the amounts of reactants shown in the equation react to give products under standard conditions. Symbol $\Delta H_r^{\ominus}$.

  Example: $2Fe(s) + 1\frac{1}{2}O_2(g) \rightarrow Fe_2O_3(s)$ $\quad \Delta H_r^{\ominus} = -824.2\,kJ\,mol^{-1}$
- **Standard enthalpy change of neutralisation:** The enthalpy change when 1 mole of water is made in the reaction between an acid and an alkali under standard conditions. Symbol $\Delta H_n^{\ominus}$.

  Example: $HCl(aq) + NaOH(aq) \rightarrow NaCl(aq) + H_2O(l)$
  $\Delta H_n^{\ominus} = -57.1\,kJ\,mol^{-1}$
- **Standard enthalpy change of solution:** The enthalpy change when 1 mole of a substance is dissolved in a very large amount of water (to infinite dilution) under standard conditions. Symbol $\Delta H_{sol}^{\ominus}$.

  Example: $NaCl(s) + aq \rightarrow Na^+(aq) + Cl^-(aq)$ $\quad \Delta H_{sol}^{\ominus} = +3.9\,kJ\,mol^{-1}$
- **Standard enthalpy change of combustion:** The enthalpy change when 1 mole of a substance is burnt in excess oxygen under standard conditions. Symbol $\Delta H_c^{\ominus}$.

  Example:
  $CH_4(g) + 2O_2(g) \rightarrow CO_2(g) + 2H_2O(l)$ $\quad \Delta H_c^{\ominus} = -890.3\,kJ\,mol^{-1}$

Finding enthalpy changes using a calorimeter

We can calculate many enthalpy changes from the results of experiments involving a **calorimeter**. A calorimeter can be a polystyrene cup, a metal can, a vacuum flask or a more complex piece of apparatus. The general procedure is:

- React known amounts of reactants in a known volume of solution.
- Measure the temperature change of the solution.
- Calculate the energy transferred using the relationship $q = mc\Delta T$
  Where q is the energy transferred in joules (J),
  m is the mass of solution in the calorimeter in grams (g),
  c is the specific heat capacity of water ($J\,g^{-1}\,°C^{-1}$). Value of $c = 4.18\,J\,g^{-1}\,°C^{-1}$,
  ΔT is the temperature change (rise or fall) (°C).
- Calculate the enthalpy change per mole of specific reactant for $\Delta H_{sol}^{\ominus}$, per mole of water formed for $\Delta H_n^{\ominus}$ or for the number of moles shown in the equation for $\Delta H_r^{\ominus}$.

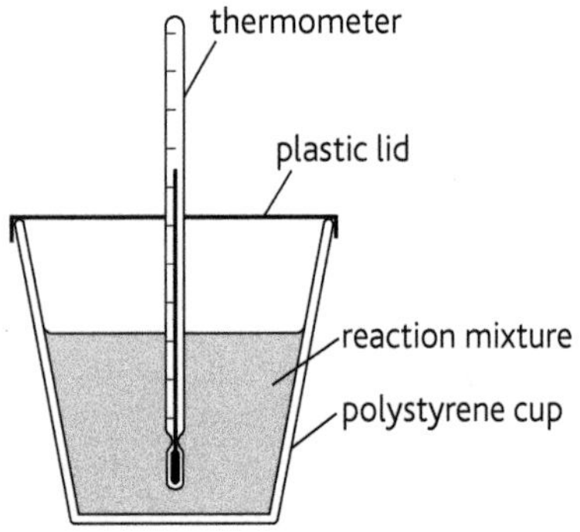

***Figure 6.3.1** A polystyrene drinking cup can act as a simple calorimeter*

$\Delta H_r^{\ominus}$ by calorimetry: worked example

0.90 g of zinc was added to 75.0 cm³ of 0.25 mol dm⁻³ copper(II) sulphate solution in a calorimeter. The copper(II) sulphate is in excess. The mixture was stirred constantly. The temperature of the solution rose from 18.0 °C to 27.9 °C. Calculate the enthalpy change for this reaction, A_r Zn = 65.4.

Step 1: Substitute the data into the equation: $q = mc\Delta T$

$$q = 75.0 \times 4.18 \times (27.9 - 18) = 3103.65\,J$$

Step 2: Calculate the energy per mole of Zn: $0.9\,g \rightarrow 3103.65\,J$

$$\text{So } 65.4\,g \rightarrow \frac{65.4}{0.9} \times 3103.65 = 225\,531.9\,J$$

Standard enthalpy change of reaction = $-230\,kJ\,mol^{-1}$ zinc (to two significant figures)

Enthalpy changes of neutralisation are found in a similar way to the above example by adding known concentrations and volumes of acid and alkali and measuring the maximum temperature rise. Enthalpy changes of solution are found by adding a known amount of solute to a known volume of water in a calorimeter and measuring the maximum rise in temperature.

Standard enthalpy change of combustion by calorimetry

A simple apparatus used to measure the enthalpy change of combustion is shown in Figure 6.3.2.

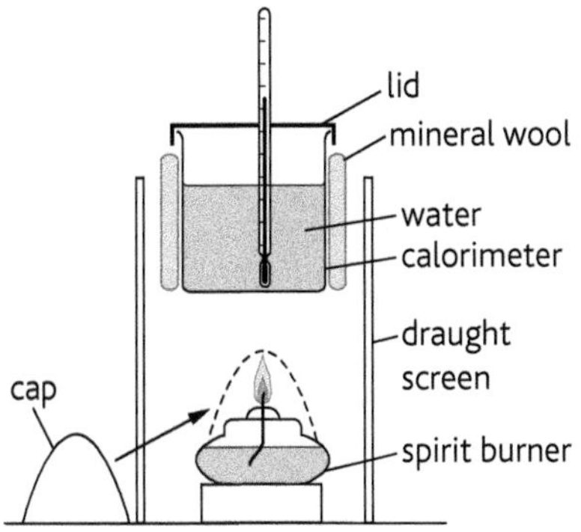

Figure 6.3.2 *Measuring the enthalpy change of combustion*

The procedure is:

- Weigh the spirit burner and cap with the fuel.
- Put a known amount of water in the calorimeter and record its temperature.
- Light the spirit burner, remove the cap and let the fuel burn until the temperature of the water in the calorimeter rises by about 10 °C.
- Remove the spirit burner, replace the cap and reweigh.

The results are processed in a similar way as for the worked example above. We need to know:

- the mass of water in the calorimeter
- the temperature rise of the water
- the mass of fuel burnt
- the molar mass of the fuel.

The value of $\Delta H_c^\ominus$ is calculated per mole of fuel burnt.

Key points

- Enthalpy change of reaction, ΔH_r, is the enthalpy change when the amounts of reactants shown in the equation react to give products.
- Enthalpy change of combustion, ΔH_c, is the enthalpy change when 1 mole of a substance is burnt in excess oxygen.
- Enthalpy change of neutralisation, ΔH_n, relates to the enthalpy change for the reaction: $H^+ + OH^- \rightarrow H_2O$
- Enthalpy change of solution, ΔH_{sol}, is the enthalpy change when 1 mole of a substance dissolves to form a very dilute solution in water.
- The symbol $^\ominus$ indicates the enthalpy change under standard conditions.
- Values for ΔH_r, ΔH_c, ΔH_n and ΔH_{sol}, can often be found by direct experiment using a calorimeter.

Exam tips

In calorimetry experiments we assume that:

- 1 cm^3 of water weighs 1 g.
- Aqueous solutions are treated as if they are water.
- The solution has the same specific heat capacity as water.
- When a solid is added to a solution, the mass of the solid is ignored.

Exam tips

You should know the main sources of error in calorimetry experiments:

- loss of heat to the air from the water / solution and from the burning fuel
- loss of heat to the calorimeter and thermometer
- evaporation of fuel.

6.4 Hess's law

Learning outcomes

On completion of this section, you should be able to:

- explain and use the term 'enthalpy change of formation'
- state Hess's law
- use Hess's law to calculate enthalpy changes of formation, reaction and combustion.

Did you know?

Hess's law is an example of the First law of Thermodynamics as applied to chemical reactions. The First law states that *Energy cannot be created or destroyed.*

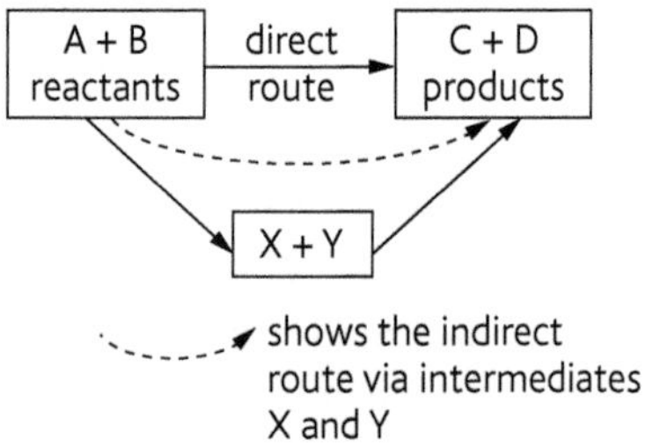

Figure 6.4.1 *The enthalpy change is the same whichever route is followed*

Enthalpy change of formation

Standard enthalpy change of formation: The enthalpy change when 1 mole of a compound is formed from its elements under standard conditions. Symbol $\Delta H_f^\ominus$.

Example: C(graphite) + $2H_2$(g) → CH_4(g) $\Delta H_f^\ominus$ [CH_4(g)] = −74.8 kJ mol^{-1}

C(graphite) is used here because graphite is the most stable form of carbon. The value of $\Delta H_f^\ominus$ of an element in its normal state at room temperature is zero.

Hess's law

Hess's law states that *the total enthalpy change for a chemical reaction is independent of the route by which the reaction proceeds.*

Figure 6.4.1 illustrates Hess's law. It shows that the enthalpy change is the same, whether you go by the direct route or the indirect route.

We can use Hess's law to calculate enthalpy changes which cannot be found directly by experiment, e.g. the enthalpy change of formation of propane.

Deducing enthalpy change of reaction using $\Delta H_f^\ominus$ values

An **enthalpy cycle** (Hess cycle) is constructed as shown in Figure 6.4.2.

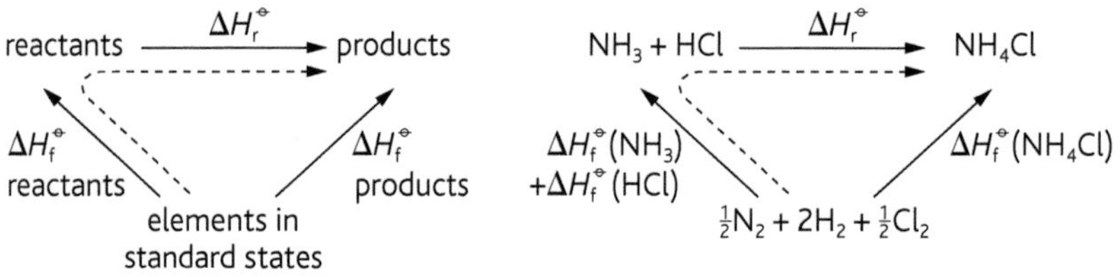

Figure 6.4.2

Using Hess's law we can see that $\Delta H_{f\ reactants}^\ominus + \Delta H_r^\ominus = \Delta H_{f\ products}^\ominus$

So $\Delta H_r^\ominus = \Delta H_{f\ products}^\ominus - \Delta H_{f\ reactants}^\ominus$

Worked example 1

Calculate $\Delta H_r^\ominus$ for the reaction: SO_2(g) + $2H_2S$(g) → 3S(s) + $2H_2O$(l)

$\Delta H_f^\ominus$ values: SO_2(g) = −296.8 kJ mol^{-1}, H_2S(g) = −20.6 kJ mol^{-1}, H_2O(l) = −285.8 kJ mol^{-1}

The stages are:

- Write the balanced equation.
- Draw the enthalpy cycle with the elements at the bottom as shown.
- Apply Hess's law.

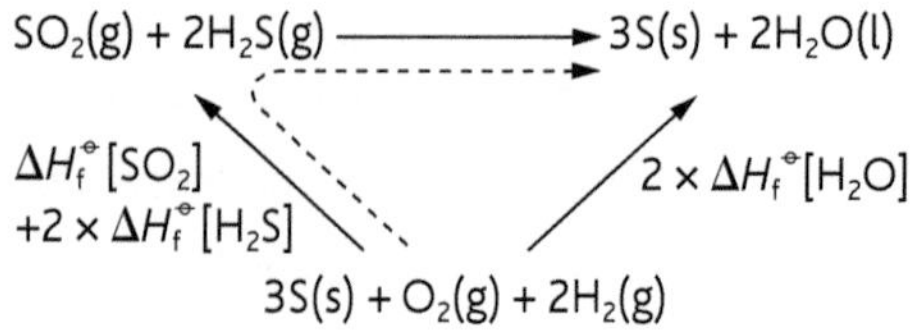

Figure 6.4.3

According to Hess's law:

$$\Delta H_f^{\ominus}[SO_2] + 2\Delta H_f^{\ominus}[H_2S] + \Delta H_r^{\ominus} = 2\Delta H_f^{\ominus}[H_2O]$$
$$-296.8 \quad + 2 \times (-20.6) + \Delta H_r^{\ominus} = 2 \times (-285.8)$$

so $\Delta H_r^{\ominus} = 2 \times (-285.8) + 296.8 + (2 \times 20.6)$
so $\Delta H_r^{\ominus} = -233.6\,kJ\,mol^{-1}$

Deducing enthalpy change of reaction using $\Delta H_c^{\ominus}$ values

The enthalpy cycle is constructed as shown below and a worked example shown.

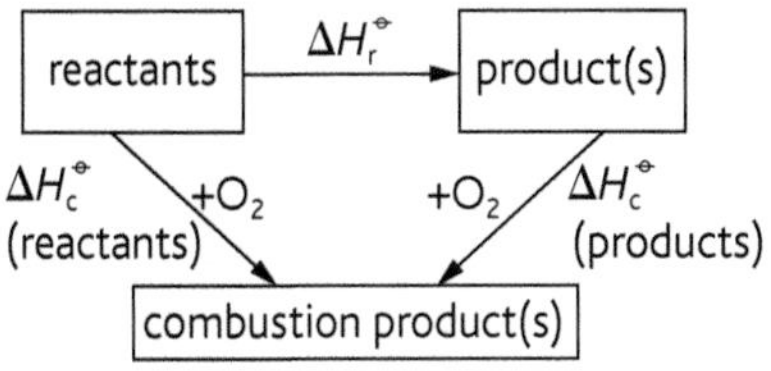

Figure 6.4.4

Worked example 2

Calculate $\Delta H_r^{\ominus}$ for the reaction: $3C(graphite) + 4H_2(g) \rightarrow C_3H_8(g)$

$\Delta H_c^{\ominus}$ values: $C(graphite) = -393.5\,kJ\,mol^{-1}$,
$H_2(g) = -285.8\,kJ\,mol^{-1}$,
$C_3H_8(g) = -2219.2\,kJ\,mol^{-1}$

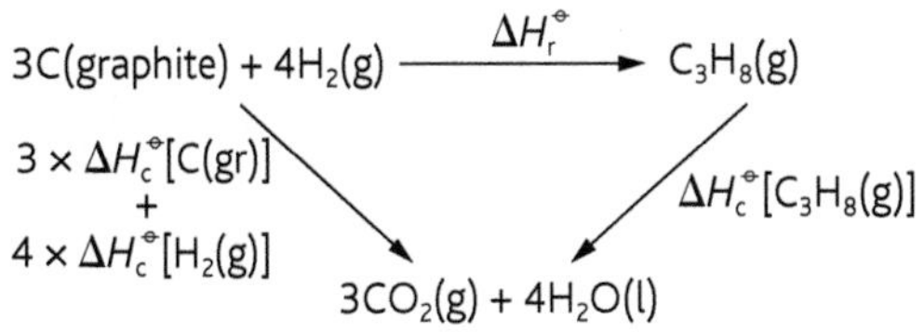

Figure 6.4.5

According to Hess's law:

$$\Delta H_r^{\ominus} + \Delta H_c^{\ominus}[C_3H_8] = 3\Delta H_c^{\ominus}[C(graphite)] + 4\Delta H_c^{\ominus}[H_2(g)]$$
$$\Delta H_r^{\ominus} + (-2219.2) = 3 \times (-393.5) + 4 \times (-285.8)$$

so $\Delta H_r^{\ominus} = 3 \times (-393.5) + 4 \times (-285.8) + 2219.2$
so $\Delta H_r^{\ominus} = -104.5\,kJ\,mol^{-1}$

Note: In this reaction, $\Delta H_r^{\ominus}$ is also the enthalpy change of formation of propane.

Exam tips

Remember that:

- $\Delta H_f^{\ominus}$ of an element is 0.
- In enthalpy cycles, make sure that you follow the arrows round in the correct direction.
- Look carefully at the state symbols when looking up enthalpy changes, e.g. $\Delta H_f^{\ominus}[H_2O(g)] = -241.8\,kJ\,mol^{-1}$ but $\Delta H_f^{\ominus}[H_2O(l)] = -285.8\,kJ\,mol^{-1}$

Key points

- Enthalpy change of formation refers to the formation of 1 mole of a compound from its elements under standard conditions.
- Hess's law states that *the total enthalpy change in a reaction is independent of the route.*
- Hess's law can be used to calculate enthalpy changes which cannot be determined directly by experiment, e.g. ΔH_r from ΔH_f or ΔH_c of reactants and products.

6.5 More enthalpy changes

Learning outcomes

On completion of this section, you should be able to:

- construct Hess cycles using bond energies
- explain the term enthalpy change of hydration, ΔH_{hyd}
- explain the term lattice energy, ΔH_{latt}
- use Hess cycles to determine ΔH_{sol} or ΔH_{hyd}.

An enthalpy cycle using bond energies

We can use the enthalpy cycle in Figure 6.5.1 to calculate an enthalpy change using bond energies. It is often easier, however, to use the 'balance sheet' method (see Section 6.2).

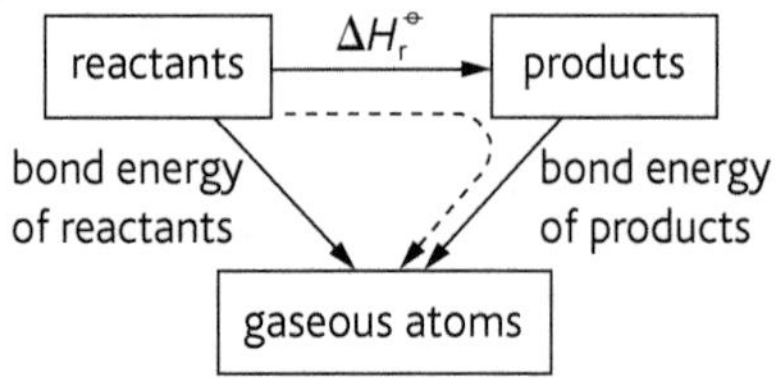

Figure 6.5.1

Worked example 1

Calculate $\Delta H_r^\ominus$ for the complete combustion of ethene to carbon dioxide and water.

E(C=C) +610, E(C–H) +410, E(O=O) +496, E(C=O) +740, E(H–O) +460

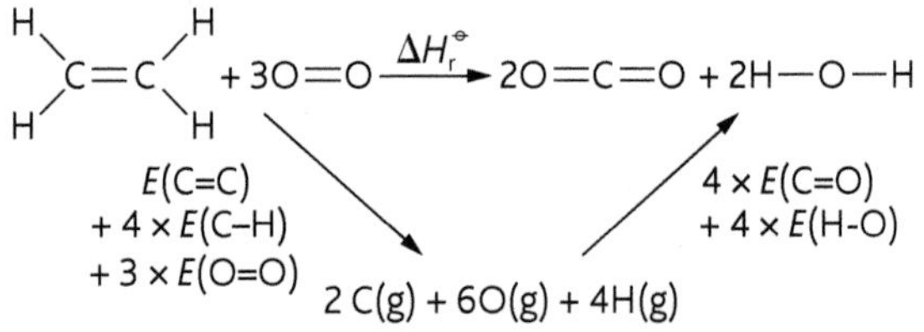

Figure 6.5.2

According to Hess's law: $\Delta H_r^\ominus$ = E(reactants) + E(products)

Note: The values of E(products) are negative because bond formation is exothermic.

E(C=C) + 4 × E(C–H) + 3 × E(O=O) + 4 × E(C=O) + 4 × E(H–O)
+ 610 + 4 × 410 + 3 × 496 + 4 × (–740) + 4 × (–460)
$\Delta H_r^\ominus = -1062\,\text{kJ mol}^{-1}$

Enthalpy change of hydration and lattice energy

Standard enthalpy change of hydration: The enthalpy change when 1 mole of a specified gaseous ion dissolves in enough water to form an infinitely dilute solution under standard conditions. Symbol $\Delta H_{hyd}^\ominus$.

Examples: $Mg^{2+}(g) + aq \rightarrow Mg^{2+}(aq)$ $\Delta H_{hyd}^\ominus = -1920\,\text{kJ mol}^{-1}$
$Cl^-(g) + aq \rightarrow Cl^-(aq)$ $\Delta H_{hyd}^\ominus = -364\,\text{kJ mol}^{-1}$

Lattice energy: The enthalpy change when 1 mole of an ionic compound is formed from its gaseous ions under standard conditions. Symbol $\Delta H_{latt}^\ominus$.

Example: $Ca^{2+}(g) + 2Cl^-(g) \rightarrow CaCl_2(s)$ $\Delta H_{latt}^\ominus = -2258\,\text{kJ mol}^{-1}$

Lattice energies are always exothermic because energy is always released when new 'bonds' are made.

Did you know?

We can calculate lattice energy in theory from the size and charge of the ions and the way they are packed. This sometimes gives slightly different values from the lattice energy obtained by experiment. 'Lattice energy' is a commonly used term and is often used to mean the same as lattice enthalpy. The difference between the two is so small that we often use the two terms to mean the same thing.

Enthalpy change of solution from $\Delta H^{\ominus}_{latt}$ and $\Delta H^{\ominus}_{hyd}$

We can calculate the enthalpy change of solution or the enthalpy change of hydration using an enthalpy cycle.

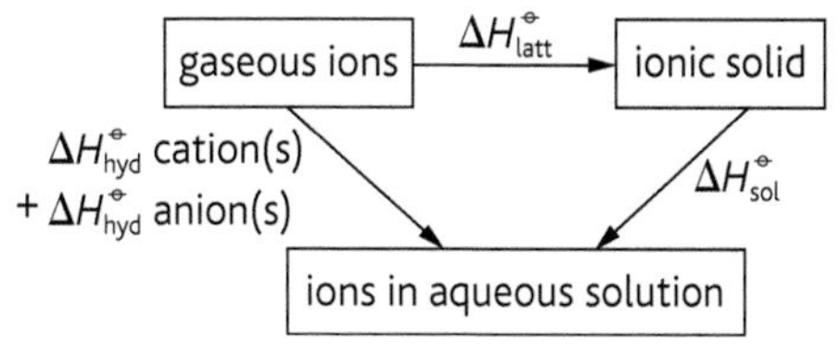

Figure 6.5.3

Here is a worked example:

Calculate the enthalpy change of solution of magnesium chloride using the following data: $\Delta H^{\ominus}_{latt}[MgCl_2] = -2592\,kJ\,mol^{-1}$, $\Delta H^{\ominus}_{hyd}[Mg^{2+}] = -1920\,kJ\,mol^{-1}$, $\Delta H^{\ominus}_{hyd}[Cl^-] = -364\,kJ\,mol^{-1}$

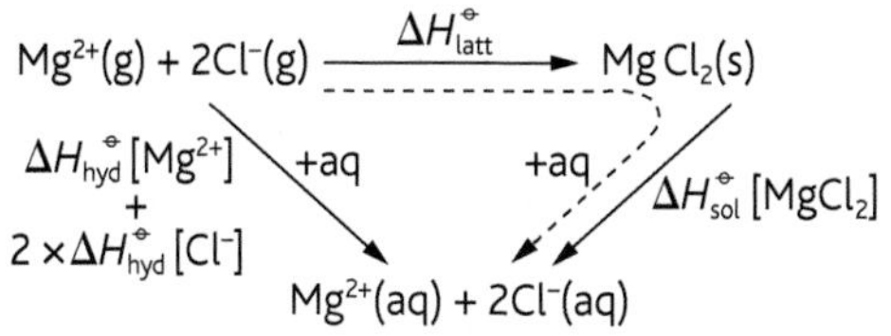

Figure 6.5.4

According to Hess's law:

$$\Delta H^{\ominus}_{latt}[MgCl_2] + \Delta H^{\ominus}_{sol}[MgCl_2] = \Delta H^{\ominus}_{hyd}[Mg^{2+}] + 2 \times \Delta H^{\ominus}_{hyd}[Cl^-]$$

So $\Delta H^{\ominus}_{sol}[MgCl_2] = \Delta H^{\ominus}_{hyd}[Mg^{2+}] + 2 \times \Delta H^{\ominus}_{hyd}[Cl^-] - \Delta H^{\ominus}_{latt}[MgCl_2]$

$$-1920 \quad + 2 \times (-364) \quad - \quad (-2592)$$

So $\Delta H^{\ominus}_{sol}[MgCl_2] = -56\,kJ\,mol^{-1}$

The value of the enthalpy change of hydration for a particular ion can also be found from this type of enthalpy cycle if we know:

- the lattice energy
- the enthalpy change of solution
- the enthalpy change of hydration of one of the ions.

Key points

- ΔH_r can be found by using a Hess cycle involving bond energies of reactants and products.
- Enthalpy change of hydration, ΔH_{hyd}, is the enthalpy change when 1 mole of gaseous ions dissolves in water to form a very dilute solution.
- Lattice energy, ΔH_{latt} is the enthalpy change when 1 mole of an ionic compound is formed from its gaseous ions.
- A Hess cycle can be drawn relating the enthalpy changes ΔH_{sol}, ΔH_{hyd} and ΔH_{latt}.

6.6 Lattice energy: the Born–Haber cycle

Learning outcomes

On completion of this section, you should be able to:

- explain the terms 'enthalpy change of atomisation' and 'electron affinity'
- calculate lattice energy from an appropriate Born–Haber cycle.

Some more enthalpy changes

In order to calculate lattice energy, we need to use ionisation energies, enthalpy change of formation and two other enthalpy changes. These two are:

Standard enthalpy change of atomisation: The enthalpy change when 1 mole of gaseous atoms is formed from an element in its standard state under standard conditions. Symbol $\Delta H^{\ominus}_{at}$.

Examples: $Na(s) \rightarrow Na(g) \quad \Delta H^{\ominus}_{at} = +107.3\,kJ\,mol^{-1}$

$\frac{1}{2}Cl_2(g) \rightarrow Cl(g) \quad \Delta H^{\ominus}_{at} = +122\,kJ\,mol^{-1}$

Values of $\Delta H^{\ominus}_{at}$ are always positive (endothermic) because energy has to be absorbed to break bonds holding the atoms together. It is important to remember that the enthalpy change of atomisation is per atom formed.

Electron affinity: The first electron affinity is the enthalpy change when 1 mole of electrons are added to a mole of gaseous atoms to form 1 mole of gaseous ions X^-. Symbol $\Delta H^{\ominus}_{ea1}$.

Examples: $Cl(g) + e^- \rightarrow Cl^-(g) \quad \Delta H^{\ominus}_{ea1} = -348\,kJ\,mol^{-1}$

$S(g) + e^- \rightarrow S^-(g) \quad \Delta H^{\ominus}_{ea1} = -200\,kJ\,mol^{-1}$

As with ionisation energies, successive electron affinities are used to form **anions** with multiple charges:

$$O(g) + e^- \rightarrow O^-(g) \quad \Delta H^{\ominus}_{ea1} = -141\,kJ\,mol^{-1}$$
$$O^-(g) + e^- \rightarrow O^{2-}(g) \quad \Delta H^{\ominus}_{ea2} = +798\,kJ\,mol^{-1}$$

Exam tips

It is important that you put the correct signs (– or +) when you carry out enthalpy change calculations to find lattice energy. Make sure you know these.

$\Delta H^{\ominus}_{ea1}$ is generally negative but $\Delta H^{\ominus}_{ea2}$ is always positive, because the incoming electron is being added to a negatively charged ion. The repulsive forces have to be overcome.

$\Delta H^{\ominus}_{i}$ values and $\Delta H^{\ominus}_{at}$ values are always positive.

$\Delta H^{\ominus}_{latt}$ are always negative.

$\Delta H^{\ominus}_{f}$ values may be positive or negative.

Born–Haber cycles

We can construct an enthalpy cycle and use Hess's law to determine the lattice energy of an ionic solid.

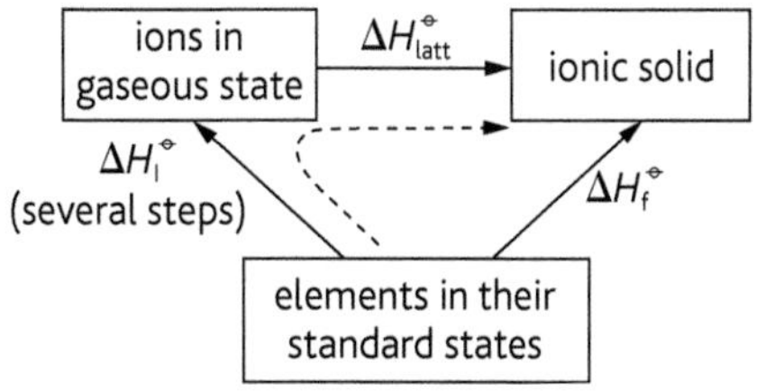

Figure 6.6.1

$\Delta H^\ominus_I$ consists of several steps:

atomise metal → atomise non-metal → ionise metal atoms → ionise non-metal atoms

So for NaBr: $\Delta H^\ominus_I = \Delta H^\ominus_{at}[Na] + \Delta H^\ominus_{at}[Br] + \Delta H^\ominus_{i1}[Na] + \Delta H^\ominus_{ea1}[Br]$

By Hess's law: $\Delta H^\ominus_I + \Delta H^\ominus_{latt} = \Delta H^\ominus_f$

So $\Delta H^\ominus_{latt} = \Delta H^\ominus_f - \Delta H^\ominus_I$

We can show all these changes on an **energy level** diagram. This is called a **Born–Haber cycle**. The Born–Haber cycle can be used to determine the lattice energy of sodium bromide as shown in Figure 6.6.2.

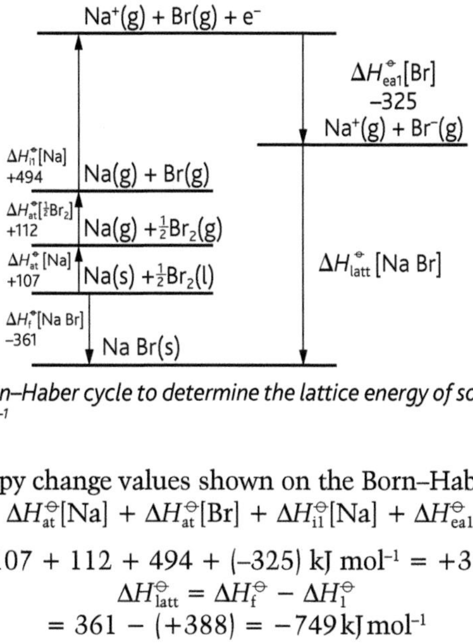

Figure 6.6.2 *A Born–Haber cycle to determine the lattice energy of sodium bromide. All figures are in kJ mol⁻¹*

From the enthalpy change values shown on the Born–Haber cycle we can see that: $\Delta H^\ominus_I = \Delta H^\ominus_{at}[Na] + \Delta H^\ominus_{at}[Br] + \Delta H^\ominus_{i1}[Na] + \Delta H^\ominus_{ea1}[Br]$

$$\Delta H^\ominus_I = +107 + 112 + 494 + (-325)\ \text{kJ mol}^{-1} = +388\ \text{kJ mol}^{-1}$$

$$\Delta H^\ominus_{latt} = \Delta H^\ominus_f - \Delta H^\ominus_I$$

$$= 361 - (+388) = -749\,\text{kJ mol}^{-1}$$

Key points

- Enthalpy change of atomisation, ΔH_{at}, is the enthalpy change when 1 mole of gaseous atoms is formed from the element.
- Electron affinity, ΔH_{ea} is the enthalpy change when 1 mole of electrons is added to 1 mole of gaseous atoms.
- A Born–Haber cycle is an enthalpy cycle which includes the enthalpy changes ΔH_{latt}, ΔH_f, ΔH_{at}, ΔH_i and ΔH_{ea}.

Exam tips

It is easier to determine lattice energy by calculating the sum of the atomisation energies, ionisation energies and electron affinities first ($\Delta H^\ominus_I$). Then apply $\Delta H^\ominus_{latt} = \Delta H^\ominus_I - \Delta H^\ominus_f$. You are less likely to make mistakes with the signs or be confused by too many enthalpy changes this way.

6.7 More about lattice energy

Learning outcomes

On completion of this section, you should be able to:

- understand how to construct Born–Haber cycles involving ions with multiple charges
- explain the effect of ionic charge and radius on the magnitude of lattice energy.

The Born–Haber cycle for magnesium oxide

Magnesium forms an ion with a 2^+ charge. The oxide ion has a charge of 2^-. So when constructing a Born–Haber cycle for magnesium oxide we have to take into account:

- the first and the second ionisation energies of magnesium and
- the first and second electron affinities of oxygen.

The enthalpy cycle and Born–Haber cycle to calculate the lattice energy of magnesium oxide are shown below. The relevant enthalpy changes (all in $kJ\,mol^{-1}$) are:

$\Delta H_f^\ominus[MgO] = -602$, $\Delta H_{at}^\ominus[Mg] = +148$, $\Delta H_{at}^\ominus[O] = +249$,
$\Delta H_{i1}^\ominus[Mg] = +736$, $\Delta H_{i2}^\ominus[Mg] = +1450$, $\Delta H_{iel}^\ominus[O] = -141$,
$\Delta H_{ea2}^\ominus[O] = +798$

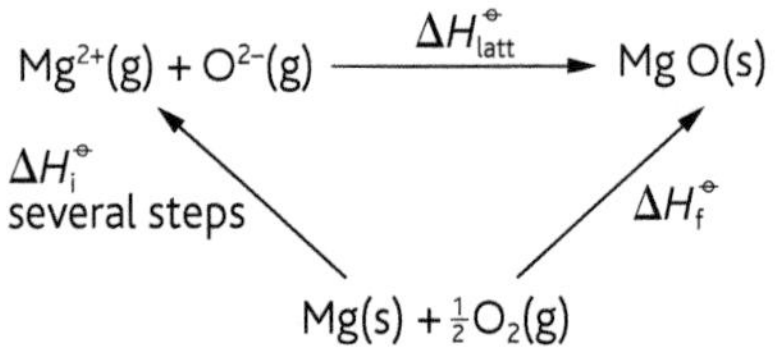

***Figure 6.7.1** Enthalpy cycle to determine the lattice energy of magnesium oxide*

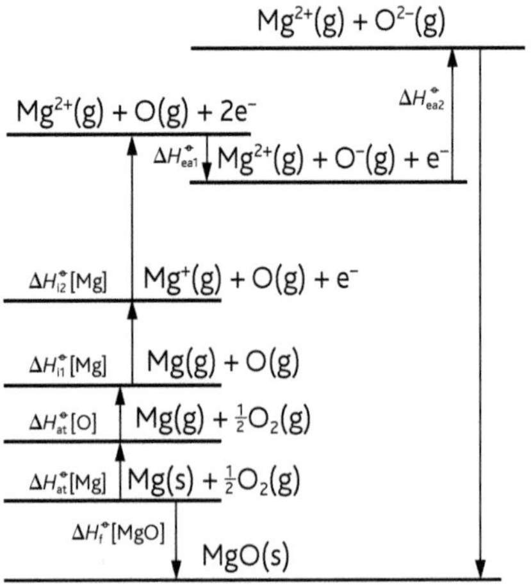

***Figure 6.7.2** Born–Haber cycle to determine the lattice energy of magnesium oxide*

From the enthalpy change values given above we can see that:

$$\Delta H_1^\ominus = \Delta H_{at}^\ominus[Mg] + \Delta H_{at}^\ominus[O] + \Delta H_{i1}^\ominus[Mg] + \Delta H_{i2}^\ominus[Mg] + \Delta H_{ea1}^\ominus[O] + \Delta H_{ea2}^\ominus[O]$$

$$\Delta H_1^\ominus = (+148) + (+249) + (+736) + (+1450) + (-141) + (+798)$$

$$= +3240\,kJ\,mol^{-1}$$

$$\Delta H_{latt}^\ominus = \Delta H_f^\ominus - \Delta H_1^\ominus$$

$$= -602 - (+3240) = -3842\,kJ\,mol^{-1}$$

Exam tips

- Take care with the signs for the first and second electron affinities. When added together they may give a positive value.
- Remember that an arrow, ↓, pointing downwards means that energy is given out and an arrow, ↑, pointing upwards means that energy is absorbed. Take care to draw these correctly if there is a second electron affinity involved.

Born–Haber cycles for $AlCl_3$, Na_2O and Al_2O_3

Aluminium chloride, $AlCl_3$, has an Al^{3+} ion and 3 Cl^- ions. So in the calculation:

- $\Delta H^{\ominus}_{at}$[Cl] and $\Delta H^{\ominus}_{ea1}$[Cl] must be multiplied by 3.
- The energy required to form Al^{3+}(g) from Al(g) is $\Delta H^{\ominus}_{i1} + \Delta H^{\ominus}_{i2} + \Delta H^{\ominus}_{i3}$

Sodium(I) oxide, Na_2O, has 2 Na^+ ions and 1 O^{2-} ion. So in the calculation:

- $\Delta H^{\ominus}_{at}$[Na] and $\Delta H^{\ominus}_{i}$[Na] must be multiplied by 2.
- The energy required to form O^{2-}(g) from O(g) is $\Delta H^{\ominus}_{ea1} + \Delta H^{\ominus}_{ea2}$

Aluminium oxide, Al_2O_3, has 2 Al^{3+} ions and 3 O^{2-} ions. So in the calculation:

- $\Delta H^{\ominus}_{at}$[Al] must be multiplied by 2 and $\Delta H^{\ominus}_{at}$[O_2] must be multiplied by 3.
- The energy required to form Al^{3+}(g) from Al(g) is $\Delta H^{\ominus}_{i1} + \Delta H^{\ominus}_{i2} + \Delta H^{\ominus}_{i3}$
- The energy required to form O^{2-}(g) from O(g) is $\Delta H^{\ominus}_{ea1} + \Delta H^{\ominus}_{ea2}$

What affects the value of lattice energies?

Lattice energy arises from the electrostatic force of attraction between oppositely charged ions. The value depends on:

- **The size of the ions.** Small ions are attracted more strongly to ions of the opposite charge than large ions. The charge density is higher on smaller ions. So, for any given anion e.g. Cl^-, the lattice energy gets less exothermic as the size of the **cation** increases:

  LiCl = $-848\,kJ\,mol^{-1}$ NaCl = $-780\,kJ\,mol^{-1}$ KCl = $-711\,kJ\,mol^{-1}$

  The same effect is seen if the cation size is kept constant:

  LiF = $-1031\,kJ\,mol^{-1}$ LiCl = $-848\,kJ\,mol^{-1}$ LiBr = $-803\,kJ\,mol^{-1}$
- **The charge on the ions.** Ions with a high charge are attracted more strongly to ions of the opposite charge than ions with a low charge. If we compare the lattice energies of LiF and MgO, which have similar sized ions, we see that the lattice energy of MgO which has doubly charged ions is very much higher than LiF which has singly charged ions:

  $\Delta H^{\ominus}_{latt}$[LiF] = $-1049\,kJ\,mol^{-1}$ $\Delta H^{\ominus}_{latt}$[MgO] = $-3923\,kJ\,mol^{-1}$
- **The way the ions are arranged in the ionic lattice.** The arrangement of the ions also affects the value of the lattice energy. It has less effect, however, than ionic charge and size.

Key points

- When constructing Born–Haber cycles involving ions with multiple charges, successive values for ΔH_i and ΔH_{ea} must be used when appropriate.
- The value of lattice energy depends on the size and charge of the ions. The lattice energy is more exothermic if the ion is smaller and has a higher charge.

Exam-style questions – Module 1

Answers to all exam-style questions can be found on the accompanying CD

Multiple-choice questions

1 What is the relative atomic mass (A_r) of bromine, given that the relative abundance of Br^{79} is 50.5% and Br^{81} is 49.50%?

A (79 × 0.495) – (81 × 0.505)
B (79 × 0.495) + (81 × 0.505)
C (79 × 0.505) – (81 × 0.495)
D (79 × 0.505) + (81 × 0.495)

2 Which of the following particles will be deflected to the greatest degree by an electric field?

A Alpha particles **C** Electrons
B Protons **D** Neutrons

3 $^{52}_{26}X$ is a radioisotope. Which of the following represents the resulting nuclide, Y, when $^{52}_{26}X$ undergoes beta decay?

A $^{52}_{27}Y$ **B** $^{52}_{28}Y$ **C** $^{48}_{24}Y$ **D** $^{56}_{28}Y$

4 Which of the following is the electronic configuration for the copper(I) ion?

A [Ar] $3d^{10}\,4s^1$ **B** [Ar] $3d^9\,4s^2$
C [Ar] $3d^9\,4s^1$ **D** [Ar] $3d^{10}$

5 Which of the following equations represents the second ionisation energy of sodium?

A $Na(g) \rightarrow Na^+(g) + e^-$ **C** $Na^+(g) \rightarrow Na^{2+}(g) + e^-$
B $Na(g) \rightarrow Na^+(g) + 2e^-$ **D** $Na^+(g) \rightarrow Na^{2+}(g) + 2e^-$

6 An element has the following five successive ionisation energies measured in $kJ\,mol^{-1}$:

1st	736	**4th**	10 500
2nd	1450	**5th**	13 600
3rd	7740		

What group of the Periodic Table is the element likely to belong to?

A V **B** IV **C** III **D** II

7 Which of the following halides of potassium would have the largest lattice energy?

A KF **B** KCl **C** KBr **D** KI

8 Which of the following describes the shape of the carbonate ion, CO_3^{2-}?

A octahedral **C** pyramidal
B trigonal planar **D** tetrahedral

9 What is the molar mass of 0.07 g of a gas which occupies a volume of 120 cm^3 at r.t.p.?
(1 mole of gas occupies a volume of 24 dm^3 at r.t.p.)

A 14 **B** 16 **C** 28 **D** 32

10 25.0 cm^3 of aqueous iron(II) sulphate required 20.5 cm^3 of 0.02 $mol\,dm^{-3}$ potassium manganate(VII) for complete reaction. The ionic equation for the reaction is:

$$5Fe^{2+}(aq) + MnO_4^-(aq) + 8H^+(aq) \rightarrow 5Fe^{3+}(aq) + Mn^{2+}(aq) + 4H_2O(l)$$

What is the concentration ($g\,mol^{-1}$) of the solution of iron(II) sulphate?
(A_r values: Fe = 56; S = 32; O = 16)

A $0.02 \times \frac{25.0}{20.5} \times \frac{1}{5} \times 152$

B $0.02 \times \frac{20.5}{25.0} \times \frac{1}{5} \times 152$

C $0.02 \times \frac{20.5}{25.0} \times 5 \times 152$

D $0.02 \times \frac{25.0}{20.5} \times 5 \times 152$

Structured questions

11 a Describe what happens in terms of loss or gain of electrons AND change in oxidation number to an oxidising agent. [2]

b A standard solution of potassium manganate(VII) can be used to find the concentration of hydrogen peroxide. The two half equations are

$$MnO_4^- + 8H^+ + 5e^- \rightarrow Mn^{2+} + 4H_2O$$
$$H_2O_2 \rightarrow O_2 + 2H^+ + 2e^-$$

i Using the half equations above, write a balanced equation to show the reaction between potassium mangangate(VII) and hydrogen peroxide. [2]

ii By making reference to the changes in oxidation numbers of the elements, show that potassium manganate(VII) is the oxidising agent and hydrogen peroxide is the reducing agent. [4]

c When aqueous bromine was added to an aqueous solution of the potassium halide, KX , a brown solution was observed.

i Write the formula of the halide ion, X. [1]

ii Using your answer in part **i** above, write a balanced equation for the reaction between aqueous bromine and KX. [2]

iii Which of the elements, Br_2 or X_2, has the greater oxidising ability? Explain your answer using an appropriate half equation. [4]

12 a State THREE assumptions of the kinetic theory as it relates to gases. [3]

b The kinetic theory assumes that all gases are ideal however this does not exist in reality.

i State the conditions under which gases deviate from ideal behaviour and explain why the deviations occur. [4]

ii Sketch a graph of PV against P showing the curves for an ideal gas and CO_2 on the same axes. [2]

c i Use the ideal gas equation to calculate the number of moles of a gas which occupies a volume of 0.072 dm^3 at −48 °C and 3.4 atm. $R = 8.31\ JK^{-1}\ mol^{-1}$ [5]

ii If the mass of the gas in part i above is 0.94 g, calculate its relative molecular mass. [1]

13 a State Hess's law. [1]

b The enthalpies of formation of carbon dioxide and water are $-393\ kJ\ mol^{-1}$ and $-286\ kJ\ mol^{-1}$ respectively.

i Use this information to draw an energy cycle diagram for the standard enthalpy of formation of ethane (C_2H_6). [2]

ii Calculate the standard enthalpy of formation of ethane. [2]

c A student was required to carry out an experiment in the lab to determine the enthalpy of solution of ammonium nitrate. The student was given a polystyrene cup, a measuring cylinder and a thermometer.

i Why was a polystyrene cup and not a beaker used in the experiment? [1]

ii State one assumption that was made when performing the calculations for this reaction. [1]

iii Given that the initial temperature of the water was 27.3 °C and the final was temperature 15.1 °C, draw an energy profile diagram, clearly showing whether the reaction was endothermic or exothermic. [4]

iv Using the information in part **iii** above, calculate the enthalpy of solution in $kJ\ mol^{-1}$ when 16.0 g of ammonium nitrate dissolves in 100 cm^3 of water. [4]

(A_r values: N = 14; H = 1; O = 16; C = 12. The enthalpies of formation of carbon dioxide and water are $-393\ kJ\ mol^{-1}$ and $-286\ kJ\ mol^{-1}$ respectively. The standard combustion of ethane is $-1560\ kJ\ mol^{-1}$).

14 Ethanedioic acid, commonly called oxalic acid is an organic acid that occurs naturally in plants and animals and is made of carbon, hydrogen and oxygen. When oxalic acid is burnt in excess oxygen it is found to contain 26.67% of C, 2.22% of H and 71.11% of O.

a Why is excess oxygen used? [1]

b i Define 'empirical formula' and 'molecular formula'. [2]

ii If the molar mass of oxalic acid is 90 $g\ mol^{-1}$, calculate its empirical formula and molecular formula. [6]

c The concentration of oxalic acid can be found by titrating it with a standard solution of potassium manganate(VII) solution. They react in a ratio of 2 moles of potassium manganate(VII) to 5 moles of oxalic acid. In one such titration it was found that it required an average volume of 21.5 cm^3 of 0.02 $mol\ dm^{-3}$ potassium manganate(VII) to react with 25.0 cm^3 of oxalic acid.

i Describe what you would observe at the end point of the reaction. [1]

ii Calculate the number of moles of oxalic acid used in the titration. [3]

iii Calculate the molar concentration ($mol\ dm^{-3}$) of the oxalic acid. [1]

iv Using the molar mass given in part ii above, calculate the mass concentration of the oxalic acid. [1]

15 The first ionisation energies of the elements in Period 3 are as follows:

Symbol of element	Na	Mg	Al	Si	P	S	Cl	Ar
First ionisation energy/ $kJ\ mol^{-1}$	496	738	578	787	1012	1000	1251	1520

a i Define the term 'first ionisation energy'. [1]

ii Using magnesium (Mg) as an example, write an equation illustrating the first ionisation energy of the element. [1]

b i Using the table, state what the general trend in the value of the first ionisation energy is as we go across the period and account for this general trend. [3]

ii Study the table and account for any irregularities in the general trend of the first ionisation energies across the period. [5]

c i Sketch a graph to show the first five successive ionisation energies of magnesium against the number of electrons removed. [2]

ii How do these values of successive ionisation energies help to identify the group to which magnesium belongs. [3]

7 Rates of reaction

7.1 Following the course of a reaction

Learning outcomes

On completion of this section, you should be able to:

- design (and carry out) suitable experiments for studying factors which affect reaction rate.

Rate of reaction

A study of the rate of reaction is called reaction kinetics. In order to study reaction kinetics, we carry out experiments to measure the rate at which reactants are used up or products are made.

$$\text{rate of reaction} = \frac{\text{change in concentration (or amount) of reactants or products}}{\text{time taken for this change}}$$

Rate of reaction is usually expressed in concentration terms as $\text{mol dm}^{-3}\,\text{s}^{-1}$.

Following the course of a reaction

The method used to follow the progress of a reaction depends on the nature of the products and reactants. The methods fall into two groups: continuous methods and sampling methods.

Continuous methods

A particular physical property of the reaction mixture is monitored over a period of time.

- *Measuring volume of gas given off*

The volume of a gas given off in a reaction can be measured at various time intervals using a gas syringe or upturned measuring cylinder initially full of water. For example in the reaction of calcium carbonate with hydrochloric acid, the volume of carbon dioxide can be measured at various times as the reaction proceeds.

$$CaCO_3(s) + 2HCl(aq) \rightarrow CaCl_2(aq) + CO_2(g) + H_2O(l)$$

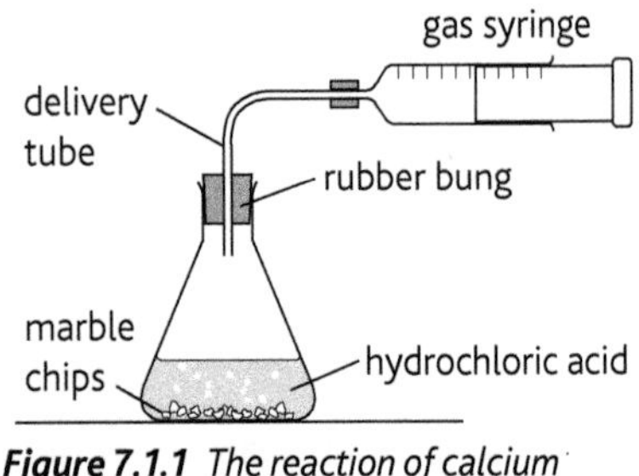

Figure 7.1.1 *The reaction of calcium carbonate with hydrochloric acid can be found by measuring the volume of carbon dioxide given off at various times*

- *Colorimetry*

This is suitable for reactions in which there is a change in the intensity of colour during the reaction. For example in the reaction between iodine and propanone in aqueous acidic solution:

$$CH_3COCH_3(aq) + I_2(aq) \rightarrow CH_3COCH_2I(aq) + HI(aq)$$

The iodine is brown in colour but the other reactants and products are colourless. As the reaction proceeds, the intensity of the brown colour fades. We can use a colorimeter to measure the amount of light absorbed by or transmitted through the reaction mixture.

A filter is chosen so that the correct wavelength of light falls on the reaction mixture in the cell. The more intense the colour of the reaction mixture, the more light is absorbed and less is transmitted through to the light-sensitive cell. As the reaction between iodine and propanone proceeds more light is transmitted through to the light-sensitive cell.

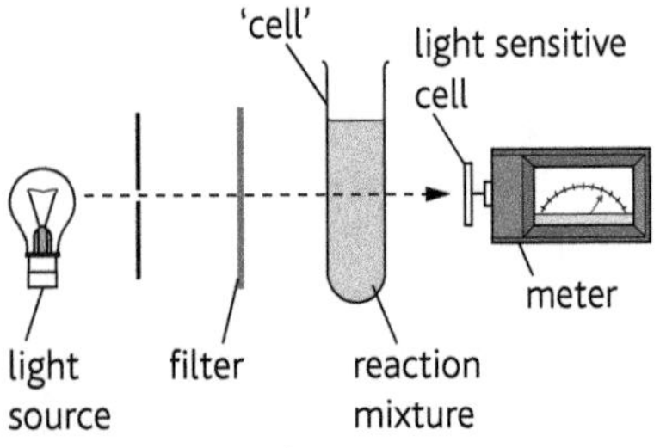

Figure 7.1.2 *A colorimeter*

Changes in electrical conductivity

If the total number of ions in a reaction mixture changes during a reaction, we can follow the reaction by measuring changes in **electrical conductivity**. We use a conductivity meter to monitor the reaction mixture. This uses alternating current so that no electrolysis of the mixture occurs. Conductivity is measured in siemens, S.

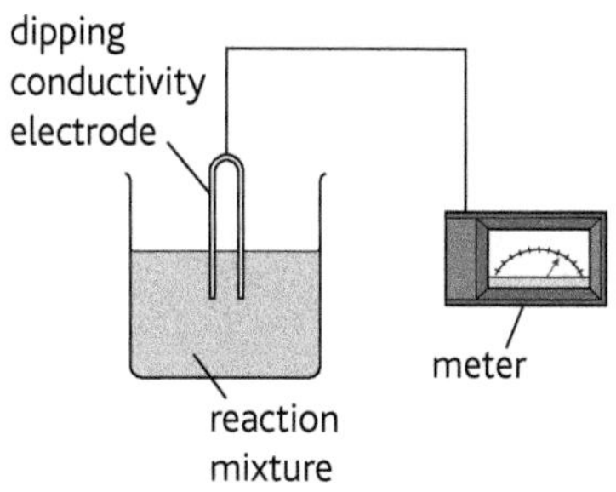

Figure 7.1.3 *A conductivity electrode connected to a meter*

An example of a reaction that can be followed by this method is:

$$(CH_3)_3CBr(l) + H_2O(l) \rightarrow (CH_3)_3COH(aq) + H^+(aq) + Br^-(aq)$$

As the reaction proceeds, the electrical conductivity of the solution increases as hydrogen ions and bromide ions are formed.

Other methods

There may be a change in pH as some reactions proceed. This may be due to H^+ ions or OH^- ions being produced or consumed. A pH meter can be use to monitor these type of reactions.

Some reactions can be followed by measuring the decrease in mass of the reaction mixture as a gas is released. This can be done by weighing, but is not very accurate unless the gas has a relatively high molar mass.

The sampling method

Small samples are removed from the reaction mixture at particular times and analysed, usually using a specific chemical reaction. Immediately after the samples are taken, they must be quenched to stop the reaction proceeding, for example:

- It is cooled rapidly so that the reaction slows down significantly.
- A chemical can be added which prevents the reaction continuing but does not affect the substance to be analysed.
- A catalyst can be removed.

An example is the reaction of propanone with iodine in acidic solution:

$$CH_3COCH_3(aq) + I_2(aq) \rightarrow CH_3COCH_2I(aq) + HI(aq)$$

A sample is taken at a particular time and added to sodium carbonate solution. Sodium carbonate reacts with the H^+ ions which catalyse the reaction. The reaction mixture can then be titrated with sodium thiosulphate to determine the concentration of iodine.

Did you know?

Hydrogen ions and hydroxide ions have very high electrical conductivities compared with other ions. This means that electrical conductivity changes can be measured even if there are other ions present as reactants or products. Hydrogen ions are really H_3O^+ ions (see Section 9.1). Hydrogen ions appear to move extremely rapidly by 'chain conduction' – they are thought to be transmitted rapidly between H_3O^+ ions and water molecules.

Key points

- The course of a reaction can be followed by measuring a particular property of a reactant or product.
- Suitable methods of following a reaction directly are measuring changes in the **i** volume of gas, **ii** electrical conductivity, **iii** colour intensity or **iv** pH.
- Some reactions can be followed by taking a sample for analysis at various time intervals as the reaction proceeds.

7.2 Calculating rate of reaction

Learning outcomes

On completion of this section, you should be able to:

- describe how reaction rate varies with time
- know how to calculate reaction rate graphically.

Reaction rate and progress of a reaction

For many reactions, reaction rate changes as the reaction proceeds. When excess magnesium reacts with dilute hydrochloric acid, the concentration of the hydrochloric acid falls and the volume of hydrogen gas given off rises:

$$Mg(s) + 2HCl(aq) \rightarrow MgCl_2(aq) + H_2(g)$$

We use square brackets to indicate molar concentration. So [HCl] means molar concentration of hydrochloric acid.

[HCl] mol dm^{-3}	0.8	0.6	0.4	0.2	0.1
Time/s	0	15	43	100	180

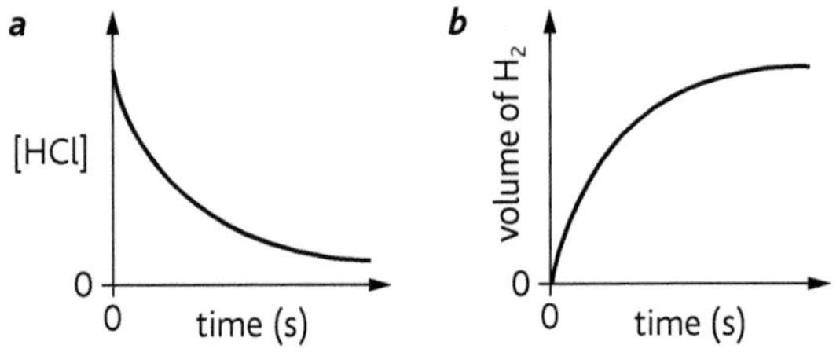

Figure 7.2.1 *The progress of the reaction beween Mg and HCl can be followed:* ***a*** *by measuring the decrease in concentration of HCl or;* ***b*** *the increase in volume of* H_2

We can see that [HCl] decreases from 0.8 to 0.6 mol dm^{-3} in the first 15 seconds.

We can write this as $\frac{\Delta\,[HCl]}{\Delta\,time} = \frac{0.8 - 0.6}{15} = 0.013\,mol\,dm^{-3}\,s^{-1}$.

The graph is a curve which gets shallower with time. So the rate is decreasing with time. If we make the time interval over which we measure the rate very small, we obtain a rate at a particular point in time i.e. at 70 s. We can calculate this rate by drawing a tangent at a particular point on the curve (see Figure 7.2.2).

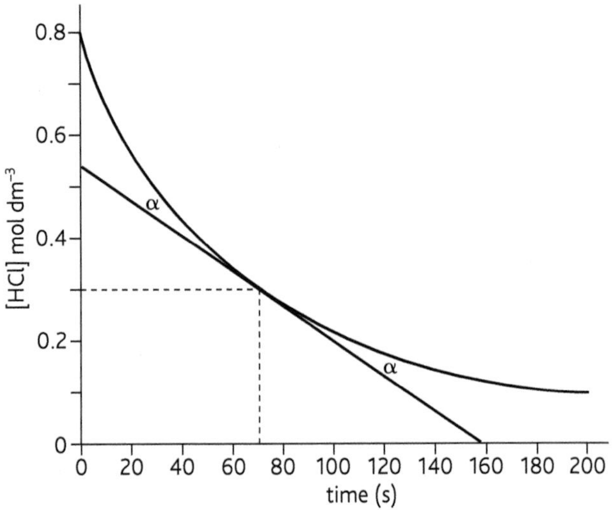

Figure 7.2.2

To calculate the rate of reaction at a particular time:

- Select a point on the line corresponding to a particular time.
- Draw a straight line at this point (the tangent) so that the two angles (α) look the same.
- Extend the tangent line to meet the axes of the graph (or at convenient values).
- Calculate the gradient of the graph. This is the rate of reaction.

In the example above, the rate of reaction at 70 s is $\frac{0.54}{157} = -3.4 \times 10^{-3}\,mol\,dm^{-3}\,s^{-1}$.

The negative sign shows that the rate is decreasing with time.

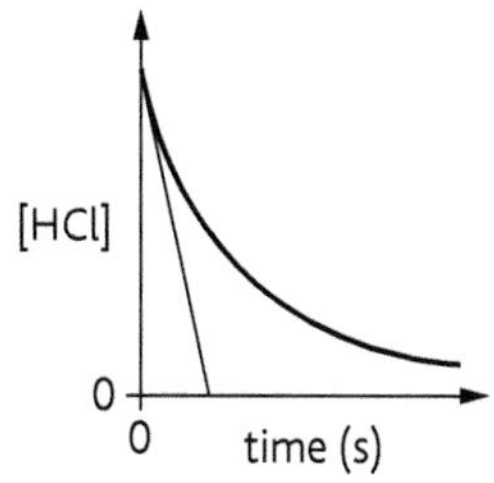

Figure 7.2.3 *Drawing a tangent to show initial rate of reaction*

Initial rate of reaction

The initial rate of reaction is calculated by drawing a tangent at the start of the curve.

Changes in rate during a reaction

We can find how reaction rate varies with the concentration of a specific reactant by drawing tangents at several different points on the graph . Looking at the gradients in Figure 7.2.4, we can see that as the reaction proceeds, the rate of reaction decreases. In Figure 7.2.4(a) we can see that each rate at a particular time corresponds to a particular concentration of hydrochloric acid.

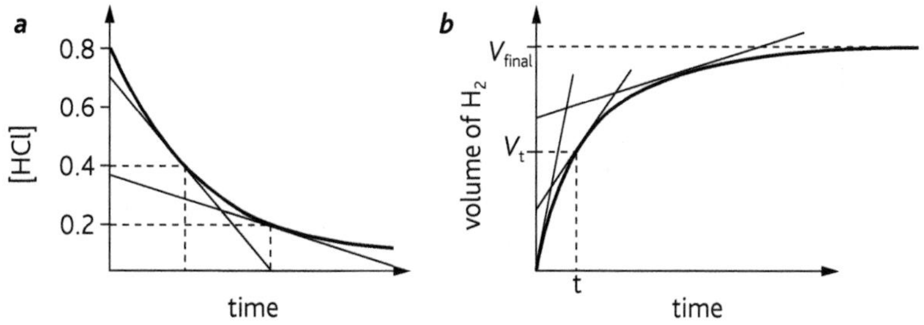

Figure 7.2.4 ***a*** *The rate of change of [HCl] and;* ***b*** *rate of volume change of H_2 both decrease with time*

In rate experiments we are interested in how the concentration of a particular reactant, rather than product, varies with time. We can use the results of experiments measuring increase in concentration or volumes of products to show what happens to the concentration of reactants. For example in Figure 7.2.4(b):

V_{final} of hydrogen – V_t of hydrogen is proportional to [HCl] at time t.

Key points

- For most reactions, rate of reaction decreases as the reaction proceeds.
- To calculate rate of reaction at specific times from a concentration–time graph, tangents are drawn at specific points on the graph (corresponding to the specific times). The rate of reaction is given by the gradient (slope) of the tangent.

7.3 Collision theory and rate of reaction

Learning outcomes

On completion of this section, you should be able to:

- explain the effect of concentration, particle size and temperature on rate of reaction
- explain the effect of catalysts, including enzymes, on the rate of reactions
- describe the role of enzymes in industrial and biological processes.

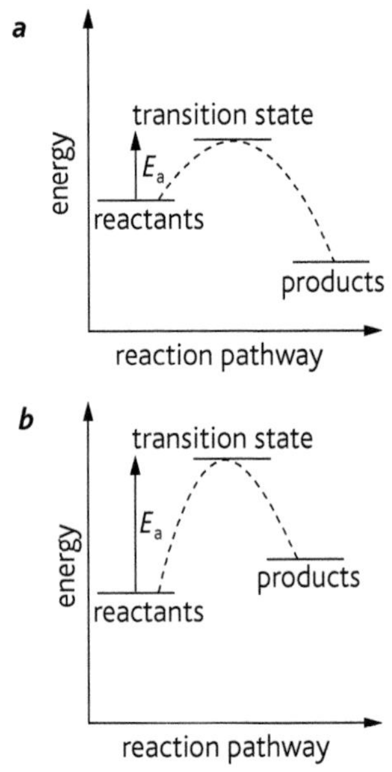

Figure 7.3.1 *Activation energy for:* ***a*** *an exothermic reaction;* ***b*** *an endothermic reaction*

Collision theory

In order to react, the reactant particles must collide with enough energy to break specific bonds in their molecules. They must also collide with the correct orientation so that the parts of each molecule that are going to react together come into contact.

The minimum energy that colliding particles must have in order to react is called the **activation energy**, symbol E_a. At this minimum energy, the molecules have a particular configuration called the **transition state**.

The effect of concentration, pressure and surface area on rate of reaction

Concentration

The more concentrated a solution, the greater is the number of particles (molecules or ions) in a given volume. There are more chances of collisions occurring. The collision frequency increases. So, rate of reaction increases.

Pressure

This is relevant for gaseous systems. The higher the pressure, the greater is the number of molecules in a given volume. There are more chances of collisions occurring. The collision frequency increases. So, rate of reaction increases.

Size of solid particles

The greater the total surface area of a solid, the greater is the number of particles that are exposed to collide with molecules in a solution or with gas molecules. If we break a large lump of solid into smaller pieces we expose a larger surface area. So, if the mass is the same, smaller particles react faster than larger particles.

Effect of temperature on rate of reaction

The Boltzmann distribution curve

The particles in any gas or solution are moving at different speeds. The energy of a particle depends on its speed. A graph of energy against the fraction of particles that have a particular energy is called a **Boltzmann distribution curve**.

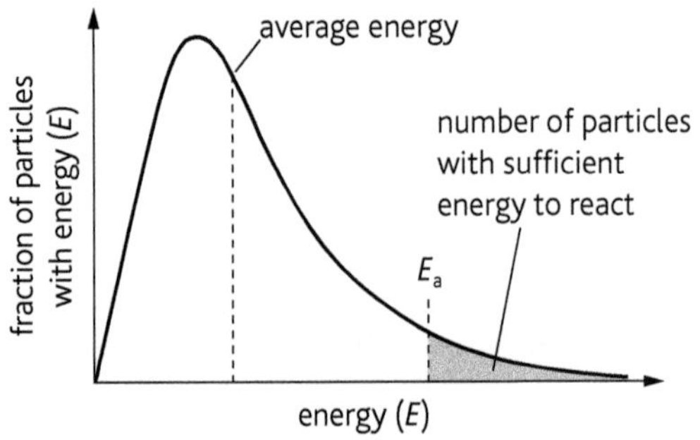

Figure 7.3.2 *Boltzmann distribution curve. The area under the graph represents the total number of particles*

The shaded area shows the proportion of molecules that have enough energy to react when they collide, i.e. equal or greater than the activation energy.

The effect of temperature on rate of reaction

Increasing the temperature changes the shape of the distribution curve slightly. An increase in temperature increases the energy of all the molecules. The average kinetic energy, however, does not increase very much. But the proportion of molecules with energy equal to or above the activation energy increases markedly (see the shaded area under the graph in Figure 7.3.3). If there are more particles with energy equal to or above the activation energy, there will be more successful collisions and the reaction rate increases.

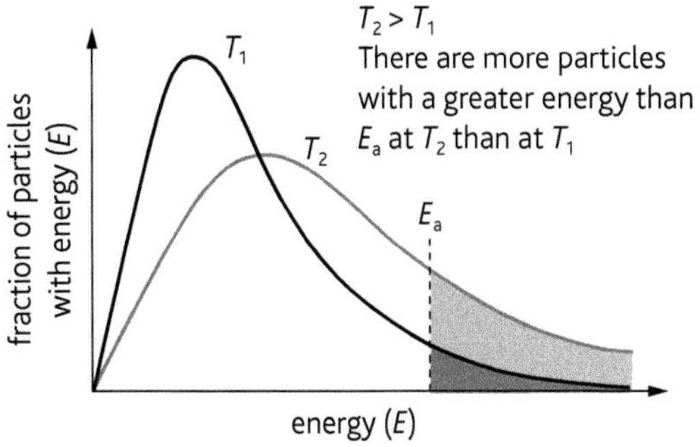

Figure 7.3.3 *Boltzmann distribution curve at a lower temperature, T_1 and a higher temperature T_2*

Catalysis

Catalysts speed up the rate of reaction by providing a different reaction path (a different mechanism) with lower activation energy. Lower activation energy leads to a greater proportion of the particles colliding successfully and therefore reacting.

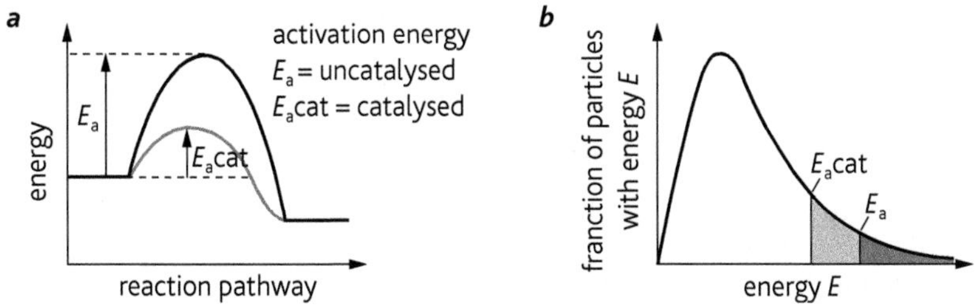

Figure 7.3.4 ***a*** *A catalyst lowers the activation energy of a reaction;* ***b*** *A greater number of particles have energies greater than or equal to the activation energy*

Enzymes

Enzymes are protein catalysts. They catalyse most of the reactions in living organisms. They have an active site on part of their surface which bonds the **substrate** and catalyses the reaction. (A substrate is a reactant which is loosely bound to the enzyme). Enzymes are very specific in the reactions they catalyse. This is because the substrate fits into the **active site** rather like a key fits a lock. This allows the bonds which are to be broken and formed to be in their in correct positions.

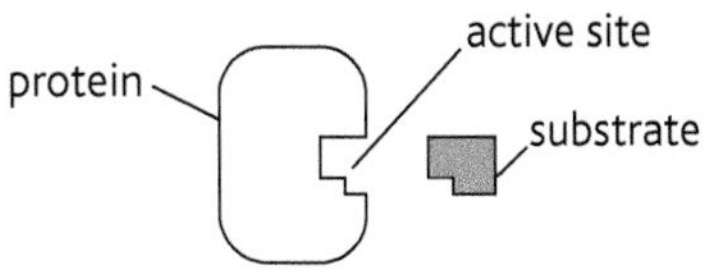

Figure 7.3.5 *The substrate fits the active site of an enzyme similar to a key fitting a lock*

Many enzymes are important in processes such as baking, brewing and the manufacture of drugs and in the manufacture of chemicals such as citric acid.

Did you know?

Most enzymes become inactive if they are left at temperatures above 40 °C for too long. The weak attractive forces holding the protein chain in its correct position are overcome and the protein changes shape. The active site is no longer in the correct conformation to catalyse the reaction. In the presence of excess substrate, many enzymes are able to function at higher temperatures. Above about 60 °C, however, the protein rapidly **denatures** because the protein chain gets tangled up with others.

Key points

- Increasing concentration of reactant (or decrease in particle size) increases reaction rate because the frequency of collisions increases.
- Increasing temperature increases reaction rate because more particles have energies greater than the activation energy.
- The Boltzmann distribution curve shows the relative number of particles having particular energies.
- Catalysts increase the rate of reaction by providing an alternative reaction pathway with lower activation energy.
- Enzymes are proteins which catalyse specific reactions.

7.4 Rate equations

Learning outcomes

On completion of this section, you should be able to:

- understand the terms 'order of reaction', 'rate equation' and 'rate constant'
- interpret concentration–time and concentration–rate graphs for zero, first and second order reactions.

Introducing rate equations

In Section 7.2, we saw how to calculate the rate of reaction at a particular time. This corresponded to a particular concentration of reactant. If we plot a graph of rate of reaction of Mg with acid against [HCl], we get a curve. But if we plot the rate of reaction against $[HCl]^2$ we get a straight line showing a proportional relationship.

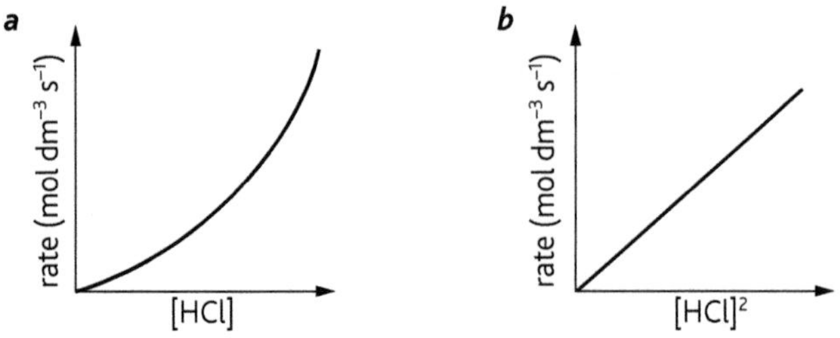

Figure 7.4.1 *Rate of reaction of HCl with Mg plotted:* ***a*** *against [HCl];* ***b*** *against* $[HCl]^2$

We can write this proportionality mathematically: rate of reaction = $k[HCl]^2$

The proportionality constant is called the **rate constant**.

The overall expression is called the **rate equation**.

The power to which the concentration of a particular reactant is raised in the rate equation (in this case 2) is called the **order** of the reaction. In this particular case, we say that the reaction is second order with respect to hydrochloric acid.

How do we deduce the rate equation?

The order of reaction cannot be determined from the stoichiometric equation. It can only be found by conducting a series of experiments. For example for the reaction:

$$2H_2(g) + 2NO(g) \rightarrow 2H_2O(g) + N_2(g)$$

We conduct experiments varying the concentration of each of the reactants one at a time.

By varying $[H_2]$ keeping [NO] constant, experiments show that rate = $k[H_2]$.

By varying [NO] keeping $[H_2]$ constant, experiments show that rate = $k[NO]^2$.

Combining these two rate equations we get the overall rate equation:

$$\text{rate} = k[H_2]\,[NO]^2$$

We say that the reaction is first order with respect to H_2, second order with respect to NO and third order overall.

Did you know?

The rate equations for some reactions may include compounds or ions not present in the chemical equation. In some instances, the order of reaction can be fractional.

How can we deduce the order of a reaction?

We can deduce the order of reaction with respect to a single reactant by:

- plotting a graph of reaction rate against concentration of reactant
- plotting a graph of concentration of reactant against time.

The shape of these graphs is characteristic.

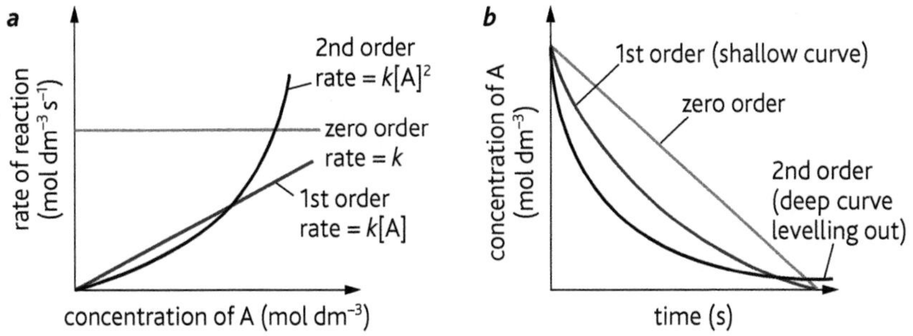

Figure 7.4.2 ***a*** *How concentration of reactant A affects reaction rate;* ***b*** *How concentration of reactant A changes with time*

Units of k

The units of k are different for different orders of reaction. For example:

For the reaction: $NO(g) + CO(g) + O_2(g) \rightarrow NO_2(g) + CO_2(g)$

the rate equation is: rate = $k[NO]^2$

Step 1: Rearrange the rate equation in terms of k:

$$k = \frac{\text{rate}}{[NO]^2}$$

Step 2: Substitute the units of rate and concentration:

$$= \frac{\text{mol dm}^{-3}\,\text{s}^{-1}}{\text{mol dm}^{-3} \times \text{mol dm}^{-3}}$$

Step 3: Cancel the units:

$$= \frac{\cancel{\text{mol dm}^{-3}}\,\text{s}^{-1}}{\cancel{\text{mol dm}^{-3}} \times \text{mol dm}^{-3}}$$

Step 4: Rearrange with the positive index first:

$$= \text{dm}^3\,\text{mol}^{-1}\,\text{s}^{-1}$$

Exam tips

- It is easy to confuse units in this section. The following facts are important:
- Units of rate are mol dm^{-3} s^{-1} (don't forget the s^{-1}).
- Units of k for zero order reactions are mol dm^{-3} s^{-1}.
- Units of k for first order reactions are s^{-1}.
- $\frac{1}{\text{mol dm}^{-3}}$ is the same as dm^3 mol^{-1}.

Key points

- Rate of reaction is related to concentration of reactants by the rate equation.
- The general form of the rate equation is rate = $k[A]^m[B]^n$ where k is the rate constant, [A] and [B] are concentrations of reactants and m and n are the orders with respect to A and B.
- Order of reaction can be determined from graphs of reaction rate against concentration or concentration against time.

Learning outcomes

On completion of this section, you should be able to:

- deduce order of reaction from appropriate data
- interpret concentration–time graphs and rate–concentration graphs
- perform calculations to find *k* or rate of reaction by substituting values in the rate equation
- understand the meaning of 'half-life'.

Order of reaction from initial rates

For a zero order reaction, a graph of concentration against time with respect to a given reactant shows a constant rate of decline (see Section 7.4). So the initial rate of reaction is given by the gradient. For first and second order reactions, the initial rate can be calculated by the tangent method from the results of several experiments using different starting concentrations of a given reagent. Figure 7.5.1(a) shows the results obtained when hydrogen peroxide at four different initial concentrations decomposes in the presence of an enzyme catalyst. The initial rates are then plotted against the initial concentrations of hydrogen peroxide (Figure 7.5.1(b)). This shows that rate = $k[H_2O_2]$. So the reaction is first order with respect to hydrogen peroxide.

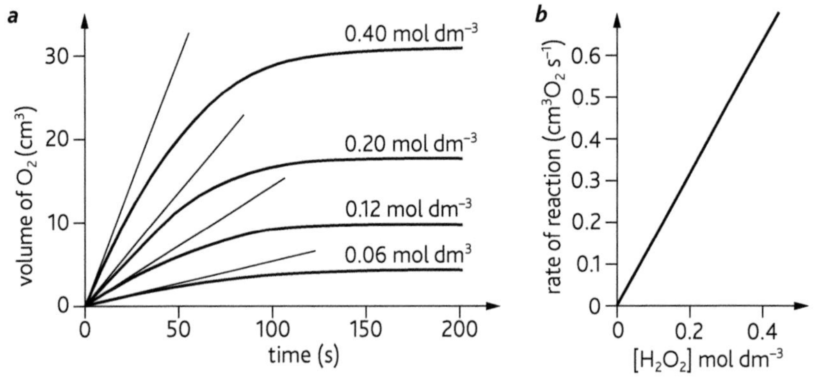

Figure 7.5.1 ***a*** *The initial rate of decomposition of* H_2O_2 *at different concentrations;* ***b*** *A plot of initial rate of decomposition of* H_2O_2 *against* $[H_2O_2]$ *shows the reaction to be first order*

Deducing a value for the rate constant

We can deduce the overall rate of reaction from limited data showing concentrations and initial rates. The table below shows data obtained from the acid catalysed **hydrolysis** of methyl methanoate:

$$HCO_2CH_3 + H_2O \xrightarrow{H^+} HCO_2H + CH_3OH$$

$[HCO_2CH_3]$/mol dm^{-3}	$[H^+]$/mol dm^{-3}	Initial rate/mol dm^{-3} s^{-1}
0.50	1.00	0.54×10^{-3}
1.00	1.00	1.10×10^{-3}
2.00	1.00	2.25×10^{-3}
2.00	2.00	4.48×10^{-3}

- Doubling the concentration of HCO_2CH_3, keeping $[H^+]$ constant, doubles the rate.
  So rate is first order with respect to $[HCO_2CH_3]$.
- Doubling the concentration of H^+ keeping $[HCO_2CH_3]$ constant also doubles the rate.
  So rate is first order with respect to $[H^+]$.

The overall rate equation is: rate = $k\,[H^+][HCO_2CH_3]$.

The rate constant can be deduced by substituting any of the sets of experimental values into the rate equation. Taking the first line of values from the table above:

$$0.54 \times 10^{-3} = k\,(0.50) \times (1.00) \text{ So } k = 1.08 \times 10^{-3}\ \text{dm}^3\,\text{mol}^{-1}\,\text{s}^{-1}$$

Half-life and order of reaction

Half-life is the time taken for the concentration of a reactant to fall to half its original value. Symbol: $t_{1/2}$.

Figure 7.5.2 shows several successive half-lives for the enzyme-catalysed decomposition of hydrogen peroxide. You can see that successive half-lives are constant. This is characteristic of a first order reaction.

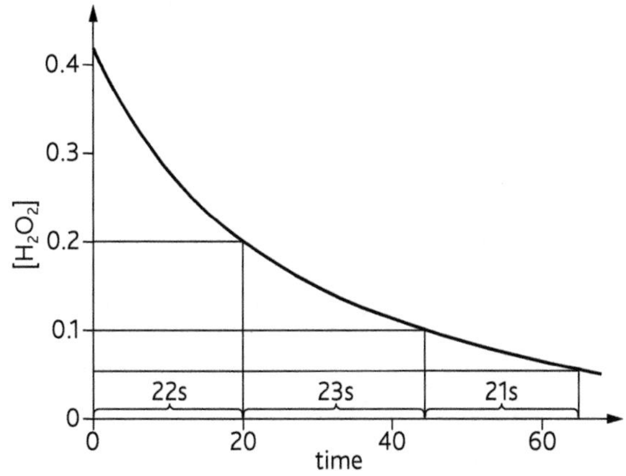

***Figure 7.5.2** The successive half-lives for the decomposition of H_2O_2 are constant (about 22 s). So the reaction is first order with respect to hydrogen peroxide*

For zero order reactions the successive half-lives decrease. For second order reactions the successive half-lives increase.

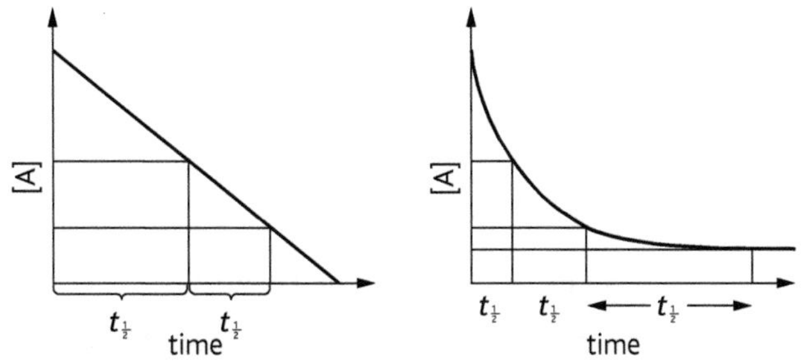

***Figure 7.5.3** **a** Successive half-lives for a zero order reaction decrease; **b** Successive half-lives for a second order reaction increase*

Did you know?

The half-life for a first order reaction may be used to find the first order rate constant using the relationship:

$$t_{1/2} = \frac{0.693}{k}$$

Key points

- Order of reaction can be found by **i** measuring initial rate in several experiments **ii** determining the change in concentration of a reactant as the reaction proceeds.
- The rate constant can be found by substituting relevant values into the rate equation.
- Half-life is the time taken for the concentration of a reactant to halve.
- In a first order reaction, half-life is independent of the concentration of reactant.

7.6 Reaction mechanisms

Learning outcomes

On completion of this section, you should be able to:

- understand the term 'rate determining step'
- deduce possible reaction mechanisms from appropriate data.

The rate determining step

Chemical reactions, especially in organic chemistry occur in a number of steps. We call this the **reaction mechanism**. Each of these steps may involve intermediates which are highly reactive. The slowest step in the reaction mechanism is called the **rate determining step**. The rate determining step controls the overall rate of reaction. So in the reaction sequence,

	Step 1		**Step 2**		**Step 3**	
Reactant A + Reactant B	→	Intermediate X	→	Intermediate Y	→	Product C
	slow		fast		fast	

The overall rate of reaction depends on Step 1, the conversion of reactants to intermediate X.

The substances that appear in the rate equation are usually those involved in the rate determining step.

Example 1: The hydrolysis of 2-bromo-2-methylpropane

Aqueous sodium hydroxide hydrolyses 2-bromo-2-methylpropane:

$$(CH_3)_3CBr + OH^- \rightarrow (CH_3)_3COH + Br^-$$

The reaction is first order with respect to 2-bromo-2-methylpropane and zero order with respect to OH^- ions. So rate $= k[(CH_3)_3CBr]$.

This suggests that the OH^- ion is not involved in the rate determining step. So $(CH_3)_3CBr$ does not react directly with OH^- ions.

Chemists think that the reaction occurs in two steps:

First step: ionisation of 2-bromo-2-methylpropane

$$(CH_3)_3CBr \xrightarrow{\text{slow}} (CH_3)_3C^+ + Br^-$$

Second step: reaction of $(CH_3)_3C^+$ with OH^-

$$(CH_3)_3C^+ + OH^- \xrightarrow{\text{fast}} (CH_3)_3COH$$

The slow step determines the overall rate of reaction. So the rate depends on the rate at which $(CH_3)_3CBr$ ionises. Hydroxide ions do not appear in the rate equation because they are involved in the rapid ionic reaction with the intermediate.

Example 2: The reaction of propanone with iodine

Propanone reacts with iodine in the presence of an acid catalyst. The catalyst provides H^+ ions for the reaction.

$$CH_3COCH_3 + I_2 \xrightarrow{H^+} CH_3COCH_2I + HI$$

The rate equation for this reaction is: rate = $k[CH_3COCH_3][H^+]$

A proposed mechanism is:

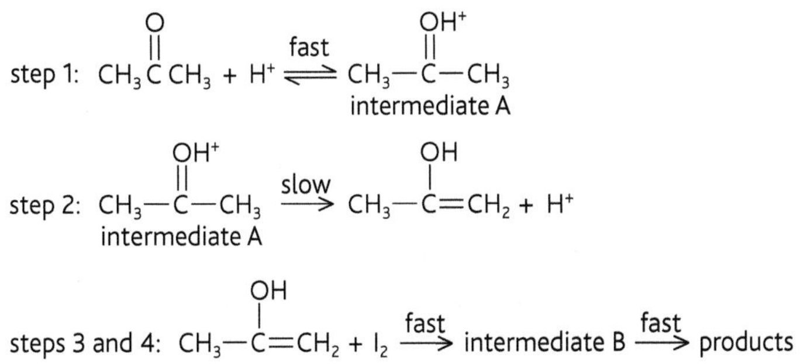

We can see that this mechanism is consistent with the rate equation. Intermediate A is derived from the CH_3COCH_3 and H^+ ions which react together to form it. This is why both CH_3COCH_3 and H^+ appear in the rate equation. Iodine is involved in a fast step later on, so does not appear in the rate equation.

Predicting order of reaction from a given reaction mechanism

If we are given a reaction mechanism, we can predict the order of reaction by reference to the rate determining step. For example, the reaction of propanone with bromine in alkaline solution. The alkaline solution contains OH^- ions. The proposed reaction mechanism is:

Step 1: $CH_3COCH_3 + OH^- \xrightarrow{\text{slow}} CH_3COCH_2^- + H_2O$

Step 2: $CH_3COCH_2^- + Br_2 \xrightarrow{\text{fast}} CH_3COCH_2Br + Br^-$

The slow step involves one molecule of CH_3COCH_3 and one OH^- ion. So the rate equation should be: rate = $k[CH_3COCH_3][OH^-]$

Key points

- The rate determining step in a reaction mechanism determines the overall rate of reaction.
- The order of reaction with respect to particular reactants shows the number of molecules involved in the rate determining step.
- The rate equation provides evidence to support a particular reaction mechanism.

Exam tips

- You need not worry about learning any of these mechanisms at present. In rate of reaction questions you will be given any relevant information about the mechanism.
- Remember that we cannot deduce the reaction mechanism directly from the rate equation. The best we can do is to see if the reaction mechanism is consistent with the rate equation.

Revision questions

1 **a** Benzenediazonium chloride decomposes above 5 °C to produce phenol, hydrochloric acid and nitrogen as shown in the following equation:

$$C_6H_5N_2^+Cl^-(aq) + H_2O(l) \rightarrow C_6H_5OH(aq) + N_2(g) + HCl(aq).$$

Briefly describe a suitable method that could be used to study the rate of this reaction.

b The equation for the reaction of iodine with propanone is:

$$I_2(aq) + CH_3COCH_3(aq) \rightarrow CH_3COCH_2I(aq) + H^+(aq) + I^-(aq).$$

Explain why a colorimetric method is suitable to monitor the rate of this reaction.

2 The questions which follow are based on the graphs shown below:

I

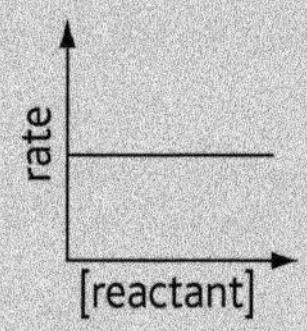

II

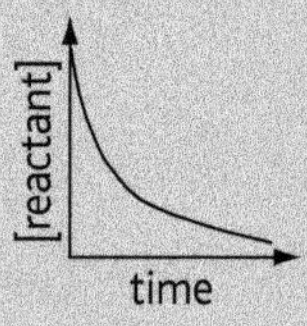

III

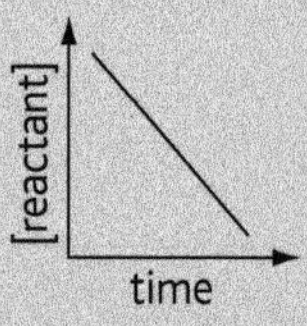

IV

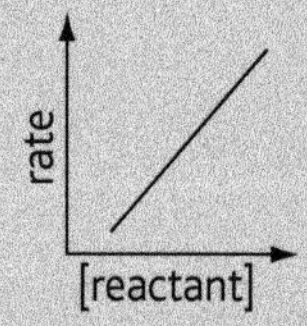

a Which of the graphs represent a zero order reaction?

b How would you determine whether Graph II represents a first or second order reaction with respect to the reactant shown?

c How would Graph IV change, if it were to represent a second order reaction, with the same labelled axes?

3 For the reaction X + Y → Z, the tabulated results below were obtained.

Experiment	[X]/ $mol\,dm^{-3}$	[Y]/ $mol\,dm^{-3}$	Initial rate/ $mol\,dm^{-3}\,s^{-1}$
1	0.030	0.015	0.1242
2	0.010	0.015	0.0138
3	0.010	0.045	0.0414

a What is the order with respect to X?

b What is the order with respect to Y?

c What is the overall order of the reaction?

d Write the rate equation for the reaction.

e Determine the value of the rate constant for this reaction.

4 The following questions are based on a student's graph shown below, which was generated from experimental data. It represents the change in concentration of a reactant [Y], over a period of 75 seconds.

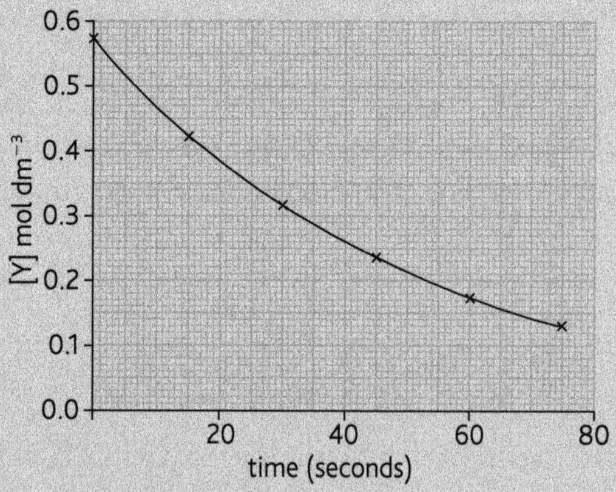

a **i** What is the initial rate?

ii What is the instantaneous rate at

- 20 seconds,
- 60 seconds?

b **i** What is meant by the term 'half-life'?

ii What are the values of the first two half-lives for this reaction?

c Giving your reasoning, state the order of the reaction with respect to reactant Y.

5 The sketch below represents the Boltzmann distribution for a sample of gas at a temperature of 500 K.

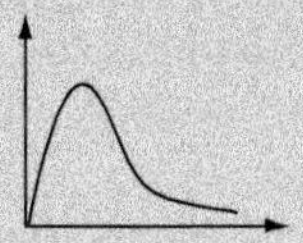

a Copy the diagram and label the axes.

b On the same axes, sketch the graph to show how the distribution would change if the temperature decreased to 390 K.

c **i** What is the effect of a decrease in temperature on the rate of a reaction?

ii Use your graphs, along with appropriate symbols and notations, to explain this effect.

6 The following questions relate to the concept of half-life ($t_{1/2}$).

a A sample of a nuclide with a half-life of 4 days is tested after 16 days.
What fraction of the sample has decayed?

A $\frac{1}{4}$
B $\frac{3}{4}$
C $\frac{7}{8}$
D $\frac{15}{16}$

b Radioactive decay is a first order process. If a newly prepared radioactive nuclide has a decay constant (rate constant) of $1 \times 10^{-6}\,s^{-1}$, what is the approximate half-life of the nuclide?

A 1 hour
B 1 day
C 1 week
D 1 month

c The half-life of Thorium-234 is 24 hours. How much of a 12 g sample would remain after 96 hours?

A 1.5 g
B 3 g
C 0.75 g
D 6 g

d A sample of a radioactive nuclide is tested 6 days after its preparation.
It is found that $\frac{7}{8}$ of the sample has decayed.
What is the half-life of this nuclide?

A 2 days
B 4 days
C 6 days
D 7 days

7 **a** Consider the reaction mechanism:
Step 1: $NO_2(g) + SO_2(g) \rightarrow NO(g) + SO_3(g)$ *slow*
Step 2: $2NO(g) + O_2(g) \rightarrow 2NO_2(g)$ *fast*

i What is the overall equation?

ii Which step is the rate-determining step?

iii What is the rate equation?

b A reaction is described by the rate equation, rate = $k[NO_2][F_2]$

i Giving your reasoning, state which of these mechanisms is consistent with the observed rate equation?

I **Step 1:** $NO_2 + F_2 \rightarrow FNO_2 + F$ *slow*
Step 2: $NO_2 + F \rightarrow FNO_2$ *fast*

II **Single step mechanism:**
$2NO_2 + F_2 \rightarrow 2FNO_2$

ii What would happen to the rate of the reaction if the concentration of F_2 were doubled, while the concentration of NO_2 remained the same?

8 Chemical equilibria

8.1 Characteristics of chemical equilibria

Learning outcomes

On completion of this section, you should be able to:

- understand the principles of chemical and physical–chemical equilibria
- state the characteristics of a system in dynamic equilibrium.

Reversible reactions

Many reactions continue until one of the reactants runs out. The reaction then stops. We say the reaction goes to completion, e.g. reacting magnesium with hydrochloric acid to form magnesium chloride and hydrogen. This reaction is irreversible.

Some reactions are reversible e.g.

$$CuSO_4{\cdot}5H_2O \underset{\text{backward reaction}}{\overset{\text{forward reaction}}{\rightleftarrows}} CuSO_4 + 5H_2O$$

Heating blue hydrated copper(II) sulphate turns it to white anhydrous copper(II) sulphate. Adding water to the anhydrous copper(II) sulphate turns it back to blue hydrated copper(II) sulphate (backward reaction).

Reversible reactions in which the forward and backward reactions take place at the same time are called **equilibrium reactions**. The products react to give the reactants at the same time as the reactants form products. These reactions do not go to completion.

To show an equilibrium reaction we use the equilibrium sign $\rightleftharpoons$. For example, the reaction of hydrogen gas and iodine vapour to produce hydrogen iodide gas:

$$H_2(g) + I_2(g) \rightleftharpoons 2HI(g)$$

We can approach equilibrium by starting off with either reactants alone or products alone. Figure 8.1.1(a) shows how the concentrations of H_2 and I_2 change when a mixture of hydrogen (0.02 mol dm^{-3}) and iodine (0.02 mol dm^{-3}) is heated in a sealed tube at 500 K. When equilibrium is reached, there is no further change in the concentration of the reactants and products. The same equilibrium is reached when we start with hydrogen iodide (0.04 mol dm^{-3}) alone (see Figure 8.1.1(b)).

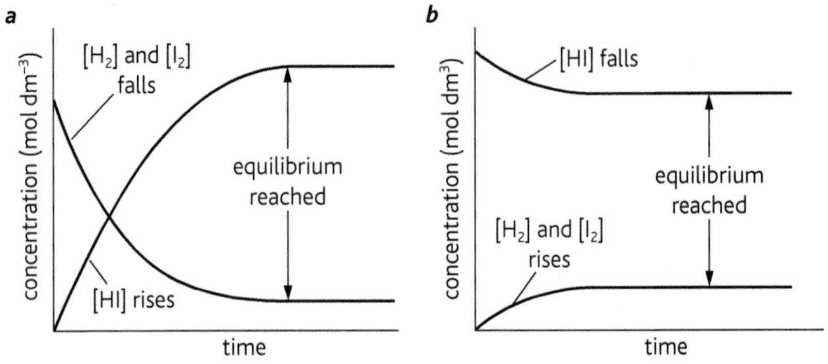

***Figure 8.1.1** The change in concentrations of reactants and products as equilibrium is approached: **a** starting with 0.02 mol dm^{-3} $H_2(g)$ and 0.02 mol dm^{-3} $I_2(g)$; **b** starting with 0.04 mol dm^{-3} $HI(g)$*

Characteristics of the equilibrium state

- It is **dynamic**: molecules of reactants are continually being converted to products and molecules of products are continually being converted to reactants.
- At equilibrium the rate of the forward and backward reactions are equal.
- The concentration of reactants and products at equilibrium do not change. They are constant. This is because the rates of the forward and backward reactions are the same.
- Equilibrium only occurs in a **closed system**. A closed system is one where none of the reactants or products escapes from the reaction mixture. Many reactions, however, can take place in open beakers if the reaction takes place in solution and no gases are involved.

Liquid–vapour equilibrium

A simple example of an equilibrium in physical chemistry is the equilibrium set up between a liquid and its vapour in a closed container.

If we put water in a closed container, water molecules with the most energy start to escape from the surface. They become a vapour. As more molecules escape, the concentration of water molecules in the vapour increases. They get closer to each other and increasing attractive forces between the molecules cause the vapor to condense. Eventually a point is reached where the molecules are moving from the liquid to the vapour at the same rate as from the vapour to the liquid. A liquid–vapour equilibrium is reached. The pressure exerted by a vapour in equilibrium with a liquid in a closed system is called its **vapour pressure**. For water, we can write this equilibrium as:

$$H_2O(l) \rightleftharpoons H_2O(g)$$

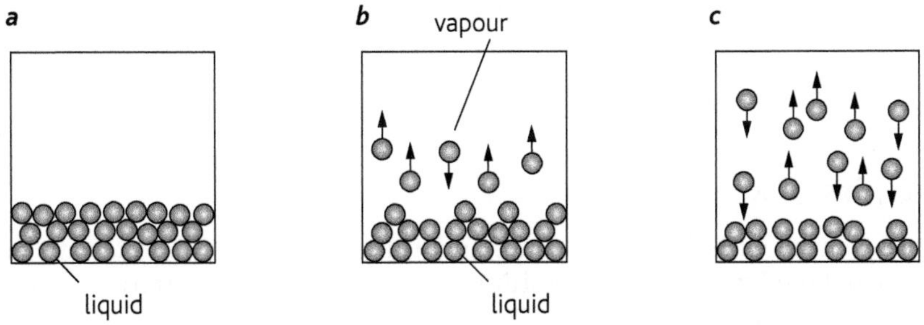

Figure 8.1.2 *a A liquid is placed in a sealed container; b Liquid molecules move from liquid to vapour at a greater rate than they return from vapour to liquid; c At equilibrium the molecules move from liquid to vapour at the some rate as they move from vapour to liquid*

Key points

- Chemical equilibrium is dynamic: the backward and forward reactions occur at the same time.
- Equilibrium is reached when the rate of forward and backward reactions are the same.
- Equilibrium only occurs in a closed system.
- In a closed system, a liquid is in equilibrium with its vapour.

Learning outcomes

On completion of this section, you should be able to:

- construct equilibrium expressions in terms of concentration
- deduce the units of K_c
- describe an experiment to determine a value for K_c.

The equilibrium constant, K_c

If we put hydrogen and iodine in a sealed tube and heat it at a constant high temperature both hydrogen and iodine are present in the gas phase, together with the product, hydrogen iodide:

$$H_2(g) + I_2(g) \rightleftharpoons 2HI(g)$$

When we measure the concentration of each reactant and product at equilibrium, we find there is a relationship between these concentrations. For this reaction the relationship is:

$$\frac{[HI]^2}{[H_2][I_2]} = \text{constant } (K_c)$$

- The constant is called the **equilibrium constant**, K_c.
- [HI] means concentration of hydrogen iodide in mol dm^{-3}.
- $_c$ in K_c refers to the equilibrium constant in terms of concentrations.
- The whole relationship is called the **equilibrium expression**.

The value of K_c is the same (at the same temperature and pressure) whatever the concentrations of hydrogen, iodine and hydrogen iodide we start with.

For any equilibrium reaction we can write an equilibrium expression which is based on the stoichiometric equation. For the equation:

$$mA + nB \rightleftharpoons pC + qD$$

The equilibrium expression is: $K_c = \frac{[C]^p[D]^q}{[A]^m[B]^n}$

Example 1: $N_2(g) + 3H_2(g) \rightleftharpoons 2NH_3(g)$ $\quad K_c = \frac{[NH_3]^2}{[N_2][H_2]^3}$

Example 2: $2SO_2(g) + O_2(g) \rightleftharpoons 2SO_3(g)$ $\quad K_c = \frac{[SO_3]^2}{[SO_2]^2[O_2]}$

Exam tips

When writing equilibrium expressions, take care to ignore any solids that appear in the stoichiometric equation. Their concentration remains constant however much there is. So the equilibrium expression for the reaction:

$$Fe_2O_3(s) + 3CO(g) \rightleftharpoons 2Fe(s) + 3CO_2(g)$$

is

$$K_c = \frac{[CO]^3}{[CO_2]^3}$$

Units of K_c

These vary with the form of the equilibrium expression. They have to be worked out by substituting mol dm^{-3} in each of the 'concentration boxes' in the equilibrium expression. For example:

$$K_c = \frac{[SO_3]^2}{[SO_2]^2[O_2]} \text{ units} = \frac{(\text{mol dm}^{-3}) \times (\text{mol dm}^{-3})}{(\text{mol dm}^{-3}) \times (\text{mol dm}^{-3}) \times (\text{mol dm}^{-3})}.$$

$$\text{So: units} = \frac{\cancel{(\text{mol dm}^{-3})} \times \cancel{(\text{mol dm}^{-3})}}{\cancel{(\text{mol dm}^{-3})} \times \cancel{(\text{mol dm}^{-3})} \times (\text{mol dm}^{-3})} = \frac{1}{\text{mol dm}^{-3}} = \text{dm}^3\,\text{mol}^{-1}.$$

If the top and bottom of the expression cancel completely then there are no units.

Determining a value of K_c by experiment

The value of K_c for the reaction

$$\underset{\text{ethanoic acid}}{CH_3COOH} + \underset{\text{ethanol}}{C_2H_5OH} \rightleftharpoons \underset{\text{ethyl ethanoate}}{CH_3COOC_2H_5} + \underset{\text{water}}{H_2O}$$

can be found using a titration method. The reaction is catalysed by an acid.

The procedure is:

- Make up mixtures of known concentration and volumes of the reactants and catalyst, e.g. 2 cm^3 water, 2 cm^3 ethanol, 1 cm^3 ethanoic acid and 5 cm^3 of 2 mol dm^{-3} HCl. (You could also start with known amounts of ethyl ethanoate, water and acid.)
- Use a burette to deliver each of the liquids into a stoppered bottle. Shake the mixture well and allow it to stand for 1 week at room temperature to reach equilibrium.
- After 1 week, titrate the whole of the mixture with 1 mol dm^{-3} sodium hydroxide using phenolphthalein as an indicator. During this time, the amount of acid in the reaction mixture decreases as the ethanoic acid reacts with the ethanol.
- Find the exact concentration of the hydrochloric acid catalyst you added by a similar titration.

Calculation of concentrations:

- Calculate number of moles of ethanoic acid, a, ethanol, b, water, c, and hydrochloric acid at the start of the experiment from the volumes and concentrations used.
- Moles of ethanoic acid at equilibrium = moles of acid calculated in the reaction mixture by titration – moles of hydrochloric acid at the start of the experiment. Call this x.
- Moles of ethyl ethanoate = moles of ethanoic acid at the start of the experiment – moles of ethanoic acid at equilibrium (a – x). This is because for every mole of ethanoic acid reacted, a mole of ethyl ethanoate is formed.
- Moles of ethanol at equilibrium = moles of ethanol at the start of the experiment – moles of ethyl ethanoate at equilibrium (b – (a – x)). This is because for every mole of ethanol reacted, a mole of ethyl ethanoate is formed.
- Moles of water at equilibrium = moles of water at start of experiment + moles of ethyl ethanoate at equilibrium (c + (a – x)). This is because for every mole of water formed, a mole of ethyl ethanoate is formed.
- The concentrations are then substituted into the equilibrium expression to find a value of K_c.

Exam tips

It is important that you write the equilibrium expression according to the equation that you have been given. The value of K_c depends on the way you write the equation. For example, if you write the equation for the reaction between H_2, I_2 and HI as $2HI \rightleftharpoons H_2 + I_2$ the equilibrium expression is:

$$kc = \frac{[H_2][I_2]}{[HI]^2}$$

Key points

- For an equilibrium reaction there is a relationship between the concentration of reactants and products and the equilibrium constant K_c.
- The general form of the equilibrium expression is $K_c = \frac{[C]^p[D]^q}{[A]^m[B]^n}$
  [A] and [B] = reactant concentrations
  [C] and [D] = product concentrations
  p, q, m, n are number of moles of relevant reactants/products in the stoichiometric equation.
- K_c can be calculated from the results of appropriate experiments.

Learning outcomes

On completion of this section, you should be able to:

- know how to perform calculations involving the equilibrium constant, K_c.

Calculating K_c from equilibrium concentrations

In some equilibrium calculations we may be given the amount, in grams or moles, present at equilibrium. When doing these calculations we first need to calculate the concentration of each of these substances.

Worked example 1

Propanone reacts with hydrogen cyanide to form an addition product:

$$CH_3COCH_3 + HCN \rightleftharpoons CH_3C(OH)(CN)CH_3$$

At equilibrium there were 0.013 mol of propanone and 0.013 mol of HCN present as well as 0.011 mol of the addition product, $CH_3C(OH)(CN)CH_3$. The volume of the reaction mixture was 500 cm^3. Calculate K_c.

Step 1: Calculate the concentration of each substance in mol dm^{-3}:

CH_3COCH_3 +	HCN	$\rightleftharpoons CH_3C(OH)(CN)CH_3$
$0.013 \times \frac{1000}{500}$	$0.013 \times \frac{1000}{500}$	$0.011 \times \frac{1000}{500}$
= 0.026 mol dm^{-3}	0.026 mol dm^{-3}	0.022 mol dm^{-3}

Step 2: Write the equilibrium expression: $K_c = \frac{[CH_3C(OH)(CN)CH_3]}{[CH_3COCH_3][HCN]}$

Step 3: Substitute the values: $K_c = \frac{0.022}{0.026 \times 0.026} = 32.5\ \text{dm}^3\,\text{mol}^{-1}$

Calculating K_c using initial concentrations and concentration of product

Worked example 2

0.29 mol of ethanoic acid and 0.25 mol of ethanol are mixed together and allowed to reach equilibrium. At equilibrium 0.18 mol of ethyl ethanoate are present. Calculate K_c.

Step 1: Write the balanced equation with the necessary information below it.

	CH_3COOH +	C_2H_5OH $\rightleftharpoons$	$CH_3COOC_2H_5$ +	H_2O
At the start:	0.29 mol	0.25 mol	0	0
At equilibrium:			0.18 mol	

Step 2: Calculate the number of moles at equilibrium:
(According to the equation, for every 1 mol of $CH_3COOC_2H_5$ formed, 1 mol of H_2O is formed.)
CH_3COOH = 0.29 – 0.18 mol = 0.11 mol (since for every 0.18 mol of $CH_3COOC_2H_5$ formed, 0.18 mol of CH_3COOH is consumed)
C_2H_5OH = 0.25 – 0.18 mol = 0.07 mol (since for every 0.18 of $CH_3COOC_2H_5$ formed, 0.18 mol of C_2H_5OH is consumed)

Step 3: Write the equilibrium expression and substitute the values.

$$K_c = \frac{[CH_3COOC_2H_5][H_2O]}{[CH_3COOH][C_2H_5OH]} \quad K_c = \frac{0.18 \times 0.18}{0.11 \times 0.07} = 4.2 \text{ (no units)}$$

Note that we do not have to know the concentrations since there are the same number of concentration terms in the numerator and denominator of the equilibrium expression.

Worked example 3

A mixture of hydrogen (2.40×10^{-2} mol) and iodine (1.38×10^{-2} mol) were put in a sealed tube and left to come to equilibrium. At equilibrium 1.20×10^{-3} mol of iodine was present. Calculate K_c.

Step 1: Write the balanced equation with the necessary information below it.

	$H_2(g)$	+	$I_2(g)$	$\rightleftharpoons 2HI(g)$
At the start:	2.40×10^{-2} mol		1.38×10^{-2} mol	0
At equilibrium:			1.20×10^{-3} mol	

Step 2: Calculate the number of moles at equilibrium:
Amount of iodine consumed
$= (1.38 \times 10^{-2}) - (0.12 \times 10^{-2}) = 1.26 \times 10^{-2}$
so, mol H_2 at equilibrium
$= (2.40 \times 10^{-2}) - (1.26 \times 10^{-2}) = 1.14 \times 10^{-2}$ mol
(1.26×10^{-2} mol I_2 were consumed so the same amount of H_2 is consumed because these have the same mole ratio in the equation.)
so, mol HI at equilibrium
$= 2 \times 1.26 \times 10^{-2}$ mol $= 2.52 \times 10^{-2}$ mol (since in the stoichiometric equation every mole of iodine which reacts produces 2 mol of HI.)

Step 3: Write the equilibrium expression and substitute the values:

$$K_c = \frac{[HI]^2}{[H_2][I_2]} \quad K_c = \frac{(2.52 \times 10^{-2})^2}{(1.14 \times 10^{-2}) \times (1.20 \times 10^{-3})} = 46.4 \text{ (no units)}$$

Key points

- Values of K_c can be deduced from the concentration of reactants and products present at equilibrium together with the equilibrium expression.
- Concentrations of specific reactants or products at equilibrium can be calculated from the equilibrium expression, knowing the value of K_c together with relevant concentration data at the start of the experiment and at equilibrium.

8.4 The equilibrium constant, K_p

Learning outcomes

On completion of this section, you should be able to:

- understand the meaning of the term 'partial pressure'
- construct equilibrium expressions in terms of partial pressures
- deduce units of K_p
- perform calculations involving K_p.

Mole fraction

For reactions involving mixtures of gases, it is easier to measure pressures than to measure concentrations. When we are dealing with mixtures of gases we need to know the fraction of a particular gas present in the mixture in terms of moles. This is called the **mole fraction**.

$$\text{mole fraction} = \frac{\text{number of moles of a particular gas}}{\text{total number of moles of gases in the mixture}}.$$

In a mixture of 0.2 mol H_2 and 0.3 mol O_2 the mole fraction of H_2 is:

$$\frac{0.2}{0.2 + 0.3} = 0.4$$

Partial pressures

The pressure exerted by the molecules of a particular gas in a mixture of gases is called the **partial pressure** of that gas. Symbol, p.

$$\text{partial pressure} = \text{total pressure} \times \text{mole fraction}$$

The partial pressures of all the gases in the mixture add up to the total pressure, P_T

$$P_T = p_1 + p_2 + p_3 \ldots$$

Example:

A mixture of gases contains 1.2 mol N_2, 0.5 mol H_2 and 0.3 mol of O_2. The total pressure is 8.0×10^5 kPa. Calculate the partial pressure of hydrogen.

$$\text{partial pressure of } H_2 = (8.0 \times 10^5) \times \frac{0.5}{1.2 + 0.5 + 0.3} = 2 \times 10^5\,\text{kPa}$$

Using the equilibrium constant, K_p

Equilibrium constants in terms of partial pressures, K_p, may be calculated in a similar way to those for K_c except that no square brackets are used.

For example, the equilibrium expression for the reaction:

$$N_2(g) + 3H_2(g) \rightleftharpoons 2NH_3(g)$$

is:

$$K_p = \frac{p^2NH_3}{pN_2 \times p^3H_2}$$

Units of K_p have to be worked out in a similar way as those for K_c. The units are calculated using pascals, Pa.

For the reaction $N_2(g) + 3H_2(g) \rightleftharpoons 2NH_3(g)$, the units are worked out from the equilibrium expression as follows:

$$\frac{(Pa)^2}{(Pa) \times (Pa)^3} \quad \text{cancelling} \quad \frac{\cancel{(Pa)} \times \cancel{(Pa)}}{\cancel{(Pa)} \times \cancel{(Pa)} \times (Pa) \times (Pa)} = \frac{1}{(Pa)^2} = Pa^{-2}$$

Exam tips

Although the standard pressure is approximately 1.01×10^5 Pa, industrial chemists often use the atmosphere (atm) as a unit of pressure, 1 atm = 1.01×10^5 Pa.

You should be prepared to do calculations using atmospheres as the base unit as well as pascals.

Worked example 1

Sulphur dioxide reacts with oxygen to form sulphur trioxide:

$$2SO_2(g) + O_2(g) \rightleftharpoons 2SO_3(g)$$

At equilibrium, the mixture contains 12.5 mol SO_2, 87.5 mol O_2 and 100 mol SO_3. The total pressure of the mixture is 1.6×10^7 Pa. Calculate K_p.

Step 1: Calculate partial pressures:

$$p_{SO_2} = \frac{12.5}{200} \times 1.6 \times 10^7 = 1.0 \times 10^6\,\text{Pa}$$

$$p_{O_2} = \frac{87.5}{200} \times 1.6 \times 10^7 = 7.0 \times 10^6\,\text{Pa}$$

$$p_{SO_3} = \frac{100}{200} \times 1.6 \times 10^7 = 8.0 \times 10^6\,\text{Pa}$$

Step 2: Write the equilibrium expression:

$$K_p = \frac{p^2SO_3}{p^2SO_2 \times pO_2}$$

Step 3: Substitute the partial pressures:

$$K_p = \frac{(8.0 \times 10^6)^2}{(1.0 \times 10^6)^2 \times (7.0 \times 10^6)} = 9.1 \times 10^{-6}\,\text{Pa}^{-1}$$

Worked example 2

2 mol of nitrogen and 6 mol of hydrogen are mixed and allowed to reach equilibrium at 680 °C and 2×10^7 Pa pressure. At equilibrium, the mixture contains 3 mol of ammonia. Calculate K_p.

$$N_2(g) + 3H_2(g) \rightleftharpoons 2NH_3(g)$$

Step 1: Find the number of moles of nitrogen and hydrogen at equilibrium:

$$\text{Moles } N_2 = 2 - 1.5 = 0.5\text{ mol}$$

(For each mol NH_3 formed $\frac{1}{2}$ mol of N_2 has been used, i.e. $3 \times \frac{1}{2}$ mol)

$$\text{Moles } H_2 = 6 - 4.5 = 1.5\text{ mol}$$

(For each 2 mol of NH_3 formed 3 mol H_2 are consumed i.e. $3 \times \frac{3}{2}$ mol)

Step 2: Calculate partial pressures:

$$p_{N_2} = 2 \times 10^7 \times 0.1 = 0.2 \times 10^7\,\text{Pa}$$
$$p_{H_2} = 2 \times 10^7 \times 0.3 = 0.6 \times 10^7\,\text{Pa}$$
$$p_{NH_3} = 2 \times 10^7 \times 0.6 = 1.2 \times 10^7\,\text{Pa}$$

Step 3: Substitute the figures:

$$K_p = \frac{p^2NH_3}{pN_2 \times p^3H_2} = \frac{(1.2 \times 10^7)^2}{(0.2 \times 10^7) \times (0.6 \times 10^7)^3}$$

$$K_p = 3.3 \times 10^{-13}\,\text{Pa}^{-2}$$

Key points

- Partial pressure = total pressure × mole fraction of gas in a mixture of gases.
- The equilibrium constant in terms of partial pressures can be deduced from partial pressures of reactants and products.
- In a reaction involving gases, the quantities of reactants or products at equilibrium can be deduced from the equilibrium expression and K_p together with relevant partial pressure data.

8.5 Le Chatelier's principle

Learning outcomes

On completion of this section, you should be able to:

- apply Le Chatelier's principle to explain the effect of change in concentration, pressure or temperature on an equilibrium reaction
- know how changes in concentration, pressure or temperature affect the values of K_c and K_p
- apply Le Chatelier's principle to the Haber Process and the Contact Process.

Position of equilibrium: Le Chatelier's principle

Position of equilibrium refers to the relative amounts of reactants and products in an equilibrium mixture.

- If the concentration of products increases relative to the reactants, we say that the position of equilibrium moves to the right.
- If the concentration of reactants increases relative to the products, we say that the position of equilibrium moves to the left.

Le Chatelier's principle: *If one or more factors that affect an equilibrium are changed, the position of equilibrium shifts in the direction which opposes the change.*

Le Chatelier's principle relates to changes in concentration, pressure and temperature.

Note: Catalysts have no effect on the position of equilibrium or the value of K_c. They only speed up the rate of reaction.

Position of equilibrium: changing concentration

The effect of changing concentrations can be seen by referring to the reaction:

$$\underset{\text{ethanoic acid}}{CH_3COOH} + \underset{\text{ethanol}}{C_2H_5OH} \rightleftharpoons \underset{\text{ethyl ethanoate}}{CH_3COOC_2H_5} + \underset{\text{water}}{H_2O}$$

- Increasing the concentration of ethanoic acid (or ethanol) shifts the position of equilibrium to the right (to oppose the added reactant(s)). More ethyl ethanoate and water are formed.
- Increasing the concentration of ethyl ethanoate (or water) shifts the position of equilibrium to the left. More ethanoic acid and ethanol are formed.
- Removing water from the reaction shifts the position of equilibrium to the right (to oppose the removal of water). More ethyl ethanoate and water are formed.

Note: Changing the concentration has no effect on the value of K_c or K_p.

Position of equilibrium: changing the pressure

Change in pressure only affects reactions where there are different numbers of gas molecules on each side of the equation. For example, there is no change in the position of equilibrium on the reaction:

$H_2(g) + I_2(g) \rightleftharpoons 2HI(g)$. This because the stoichiometric equation shows 2 moles of gas on the left and 2 moles of gas on the right.

A change in pressure does, however, change the position of equilibrium in a reaction such as:

$$N_2(g) + 3H_2(g) \rightleftharpoons 2NH_3(g)$$

- Increase in pressure shifts the position of equilibrium to the right where there are fewer gas molecules (2 moles on the right and 4 moles on the left). This opposes the increase in the number of gas molecules per unit volume when the pressure increases.
- Decrease in pressure makes the molecules move further apart (their concentration decreases). The position of equilibrium shifts to the left where the gas molecules are more concentrated.

Note: Changing the pressure has no effect on the value of K_c or K_p.

Exam tips

It is a common error to suggest that a change in concentration or pressure affects the value of K_c or K_p. If we add more reactants, more products are formed until the value of K_c is restored. Remember it is only a change in temperature that changes the equilibrium constant. Remember TOCK (Temperature Only Changes K_c).

Position of equilibrium: changing temperature

A change in temperature changes the value of K_c or K_p.

- For an endothermic reaction, increase in temperature increases the value of K_c or K_p. So the position of equilibrium shifts to the right. More products are formed. So for the reaction:

$$2HI(g) \rightleftharpoons H_2(g) + I_2(g) \quad \Delta H_r^\ominus = +9.6\,kJ\,mol^{-1}$$

  increasing the temperature shifts the position of equilibrium in favour of the products and more hydrogen iodide decomposes.

- For an exothermic reaction, increase in temperature decreases the value of K_c or K_p. So the position of equilibrium shifts to the left. More reactants are formed. So for the reaction:

$$2SO_2(g) + O_2(g) \rightleftharpoons 2SO_3(g) \quad \Delta H_r^\ominus = -197\,kJ\,mol^{-1}$$

  increasing the temperature shifts the position of equilibrium in favour of sulphur dioxide and oxygen.

Did you know?

We can use Le Chatelier's principle to understand the effect of temperature on the position of equilibrium.

An increase in temperature increases the energy of the particles.

So the reaction goes in the direction of opposing the increase in energy, i.e. in the direction of energy absorbed.

Energy is absorbed in endothermic reactions.

Maximising the yield in industrial processes

Ammonia is synthesised in the Haber Process:

$$N_2(g) + 3H_2(g) \rightleftharpoons 2NH_3(g) \quad \Delta H_r^\ominus = -92\,kJ\,mol^{-1}$$

The yield can be increased by:

- Increasing the pressure: more product (NH_3) is formed because the position of equilibrium shifts in favour of fewer molecules.
- Decreasing the temperature: increases the yield of NH_3 because the reaction is exothermic.
- Removing ammonia by condensing it: the position of equilibrium shifts to the right in favour of fewer molecules.
- Using a catalyst. This speeds up the reaction rate but has no effect on the equilibrium yield.

We use compromise conditions for the synthesis of ammonia because although rate of reaction increases as temperature increases, the yield of ammonia decreases. So a temperature of about 420–50 °C is used to give a reasonable yield at a fast enough rate.

The key reaction in the Contact Process for the manufacture of sulphuric acid is:

$$2SO_2(g) + O_2(g) \rightleftharpoons 2SO_3(g) \quad \Delta H_r^\ominus = -197\,kJ\,mol^{-1}$$

We can increase the yield of SO_3 by decreasing the temperature, increasing the pressure and using a catalyst. The theory behind this is the same as for the Haber Process.

Did you know?

In the Contact Process, a high pressure is not used because:

- The yield of SO_3 produced for SO_2 is fairly high at fairly low temperatures.
- Highly corrosive sulphur oxides are less easily contained in the reactor vessel at high pressure.

Key points

- Le Chatelier's principle states: *When the conditions in a chemical equilibrium change, the position of equilibrium shifts to oppose the change.*
- Increase in concentration of reactants shifts the position of equilibrium to the right. Decrease in concentration of reactants, shifts the position to the left.
- A change in temperature affects the value of K_c or K_p but changes in concentration or pressure have no effect on these values.
- The conditions used in the Haber Process and Contact Process are chosen to get the best yield of product at a reasonable rate of reaction.

8.6 Solubility product, K_{sp}

Learning outcomes

On completion of this section, you should be able to:

- construct equilibrium expressions for solubility product, K_{sp}
- deduce units for K_{sp}
- perform calculations involving solubility product.

Solubility

Many ionic compounds dissolve in water but some are only slightly soluble or appear to be insoluble. But even 'insoluble' ionic compounds form some ions in solution to a very small extent. We say that a solution is saturated when no more solute will dissolve in it. Solubility is measured in mol dm^{-3} (or sometimes in g/100 g water).

- Potassium chloride is soluble – a saturated solution contains 35.9 g/100 g water.
- Calcium hydroxide is described as sparingly soluble – a saturated solution contains 0.113 g/100 g water.
- Silver chloride is regarded as 'insoluble'. It is in fact sparingly soluble – a saturated solution contains 1.93×10^{-4} g/100 g water.

Solubility product

When a sparingly soluble or 'insoluble' salt is added to water, an equilibrium is established between the ions in solution and the ions in the solid. The ions move from the solid to solution at the same rate as they move from solution to solid.

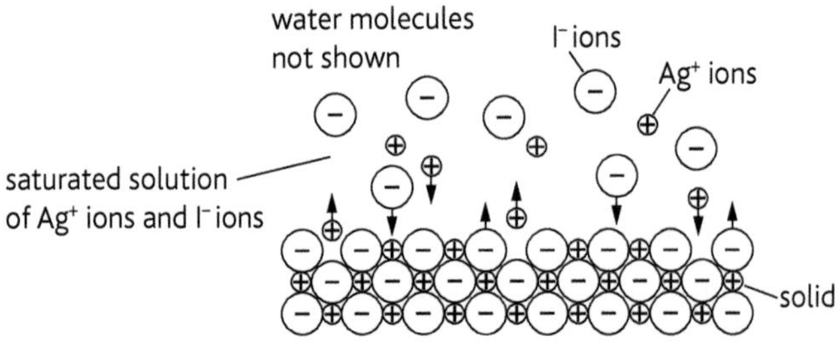

Figure 8.6.1 *At equilibrium ions move from saturated silver iodide solution to silver iodide solid at the same rate as they move from solid to solution*

We can write the equilibrium in the diagram above as:

$$AgI(s) \rightleftharpoons Ag^+(aq) + I^-(aq)$$

The related equilibrium expression is: $K = \dfrac{[Ag^+(aq)][I^-(aq)]}{[AgI(s)]}$

But the concentration of solid is constant, so we can write the expression as:

$$K_{sp} = [Ag^+][I^-]$$

K_{sp} is called the **solubility product**. The square brackets show solubility in mol dm^{-3}.

Solubility product is the product of the concentrations of each ion in a saturated solution of a sparingly soluble salt at 298 K raised to the power of their relative concentrations.

Example 1: $Co(OH)_2$ ion ratio Co^{2+} and $2OH^-$ $K_{sp} = [Co^{2+}][OH^-]^2$

Example 2: Al_2O_3 ion ratio $2Al^{3+}$ and $3O^{2-}$ $K_{sp} = [Al^{3+}]^2[O^{2-}]^3$

Units of solubility product

We calculate the units for K_{sp} in the same way as for general equilibrium expressions involving K_c.

Example:
$K_{sp} = [Co^{2+}][OH^-]^2$ units are $(mol\,dm^{-3}) \times (mol\,dm^{-3})^2 = mol^3\,dm^{-9}$

Solubility product calculations

Worked example 1: Solubility product from solubility

A saturated solution of lead iodide, PbI_2, contains 0.076 g PbI_2 in 100 g solution. Calculate K_{sp} for lead iodide. Molar mass of lead iodide = 461.0 g mol^{-1}

Step 1: Calculate the concentration of the solution in mol dm^{-3}

$$\frac{0.076}{461.0} \times \frac{1000}{100} = 1.65 \times 10^{-3}\,mol\,dm^{-3}$$

Step 2: Calculate the concentration of each ion in solution.

$$[Pb^{2+}] = 1.65 \times 10^{-3}\,mol\,dm^{-3}$$
$$[I^-] = 2 \times (1.65 \times 10^{-3})\,mol\,dm^{-3}$$
$$[I^-] = 3.30 \times 10^{-3}\,mol\,dm^{-3}$$

Step 3: Write the equilibrium expression: $K_{sp} = [Pb^{2+}][I^-]^2$

Step 4: Substitute the values: $K_{sp} = (1.65 \times 10^{-3}) \times (3.30 \times 10^{-3})^2$
$= 1.80 \times 10^{-8}$

Step 5: Add the units: $K_{sp} = 1.80 \times 10^{-8}\,mol^3\,dm^{-9}$

Worked example 2: Solubility from solubility product

Calculate the solubility in mol dm^{-3} of calcium sulphate, $CaSO_4$. $K_{sp} = 2.0 \times 10^{-5}\,mol^2\,dm^{-6}$.

Step 1: Write down the equilibrium expression:

$$K_{sp} = [Ca^{2+}][SO_4^{2-}]$$

Step 2: Rewrite the equilibrium expression in terms of one ion only:
In $CaSO_4$, $[Ca^{2+}] = [SO_4^{2-}]$ so $K_{sp} = [Ca^{2+}]^2$

Step 3: Substitute the value:

$$2.0 \times 10^{-5} = [Ca^{2+}]^2$$

Step 4: Complete the calculation and add units:

$$[Ca^{2+}] = \sqrt{2.0 \times 10^{-5}} = 4.47 \times 10^{-3}\,mol\,dm^{-3}$$

Key points

- Solubility product is an expression showing the equilibrium concentration of ions in a saturated solution of a sparingly soluble salt. It takes into account the relative number of each ion in solution.
- The value of K_{sp} can be calculated using the relevant equilibrium expression and concentration of ions present at equilibrium.
- The solubility of a sparingly soluble salt can be calculated using the value of K_{sp} and the formula of the salt.

8.7 Solubility product and the common ion effect

Learning outcomes

On completion of this section, you should be able to:

- explain the principles underlying the common ion effect
- relate solubility product to the selective precipitation of ions
- know how to determine solubility product by experiment.

Predicting precipitation

Solubility product can be used to predict whether a precipitate will form when two solutions are mixed. For example: will we get a precipitate when we mix solutions of the soluble salts barium chloride and sodium sulphate? The possible reaction is:

$$BaCl_2 + Na_2SO_4 \rightarrow BaSO_4 + 2NaCl$$

Sodium chloride is very soluble in water but barium sulphate is sparingly soluble.

For barium sulphate $K_{sp} = [Ba^{2+}][SO_4^{2-}] = 1.0 \times 10^{-10}\,mol^2\,dm^{-6}$

- If the ionic product $[Ba^{2+}][SO_4^{2-}]$ is greater than $1.0 \times 10^{-10}\,mol^2\,dm^{-6}$ a precipitate of barium sulphate will form.
- If the ionic product $[Ba^{2+}][SO_4^{2-}]$ is less than $1.0 \times 10^{-10}\,mol^2\,dm^{-6}$ barium sulphate will remain in solution. It will not precipitate.

Worked example 1

Will a precipitate form when an aqueous solution containing $6.00 \times 10^{-5}\,mol\,dm^{-3}$ barium chloride is added to an equal volume of $1.20 \times 10^{-5}\,mol\,dm^{-3}$ sodium sulphate?

$$K_{sp} \text{ of } BaSO_4 = 1.0 \times 10^{-10}\,mol^2\,dm^{-6}$$

Step 1: Calculate the concentrations: $[Ba^{2+}] = 3.00 \times 10^{-5}\,mol\,dm^{-3}$
$[SO_4^{2-}] = 6.00 \times 10^{-6}\,mol\,dm^{-3}$

Step 2: Write the equilibrium expression: $K_{sp} = [Ba^{2+}][SO_4^{2-}]$

Step 3: Substitute the figures to give the ionic product:
$= (3.00 \times 10^{-5}) \times (6.00 \times 10^{-6})$
$= 1.80 \times 10^{-10}\,mol^2\,dm^{-6}$

Step 4: Compare values of ionic product and K_{sp}:

$$1.80 \times 10^{-10} \text{ is greater than } 1.0 \times 10^{-10}$$

so a precipitate will form.

Worked example 2

What concentration of silver ions is needed in solution to precipitate silver chloride from a solution of sodium chloride containing $1.0 \times 10^{-4}\,mol\,dm^{-3}$?

$$K_{sp} \text{ of } AgCl = 1.8 \times 10^{-10}\,mol^2\,dm^{-6}$$

Step 1: Write the equilibrium expression: $K_{sp} = [Ag^+][Cl^-]$

Step 2: Substitute the figures: $(1.80 \times 10^{-10}) = [Ag^+] \times (1.0 \times 10^{-4})$

Step 3: Rearrange to find $[Ag^+]$:

$$[Ag^+] = \frac{1.80 \times 10^{-10}}{1.0 \times 10^{-4}} = 1.8 \times 10^{-6}\,mol\,dm^{-3}$$

is the minimum concentration of Ag^+ needed to cause precipitation.

Precipitation and qualitative analysis

We can use solubility product data to choose suitable reagents to selectively precipitate compounds from their solutions. For example we use silver nitrate to test for halide ions such as Cl^-, Br^- and I^-. Most halides are soluble in water as are most nitrates. But silver halides have a very low solubility product. So, adding only a few drops of silver nitrate to a halide ion results in a precipitate e.g.

$$Ag^+(aq) + Cl^-(aq) \rightarrow AgCl(s)$$

These precipitates can be distinguished by their characteristic colours.

Determining solubility product by experiment

The method is based on making a saturated solution of a particular substance and analysing the solution to find the concentration of a particular ion which is present in the solid as well. For example to find K_{sp} for calcium hydroxide, $Ca(OH)_2$:

- Add enough calcium hydroxide to a known volume of water so that solid is present as well as a solution. Shake and leave for 24 hours to reach equilibrium.
- Filter off the solid calcium hydroxide.
- Titrate samples of the calcium hydroxide solution with hydrochloric acid of known concentration.
- Calculate the concentration of the calcium hydroxide solution from the titration results.
- Find K_{sp} by substituting in the equilibrium expression $K_{sp} = [Ca^{2+}][OH^-]^2$.

The common ion effect

The **common ion effect** is the reduction in the solubility of a dissolved salt caused by adding a solution of a compound which has an ion in common with the dissolved salt. The solubility of $BaSO_4$ in water is $1.0 \times 10^{-5}\,mol\,dm^{-3}$. But the solubility of $BaSO_4$ in $0.10\,mol\,dm^{-3}$ sulphuric acid, H_2SO_4, is only $1.0 \times 10^{-9}\,mol\,dm^{-3}$. We can explain this difference by referring to equilibrium expression:

$$K_{sp} = [Ba^{2+}][SO_4^{2-}] = 1.0 \times 10^{-10}\,mol^2\,dm^{-3}$$

The concentration of SO_4^{2-} ions in the solution $= 0.10\,mol\,dm^{-3}$ (ignoring the very small concentration of SO_4^{2-} ions from barium sulphate).

Substituting the concentration values: $1.0 \times 10^{-10} = [Ba^{2+}] \times 0.1$

So $[Ba^{2+}] = 1.0 \times 10^{-9}\,mol\,dm^{-3}$ = solubility of barium sulphate in H_2SO_4

The common ion effect and practical chemistry

In quantitative work involving preparing and weighing solids (gravimetric analysis), it is important that no solid is lost. The common ion effect can help here. For example: the concentration of sulphate in aqueous solution can be determined gravimetrically by adding barium chloride. This precipitates barium sulphate. If a precipitate of barium sulphate in water is washed with water, some barium and sulphate ions will dissolve and will be lost in the filtrate. So the precipitate is filtered and then washed with dilute sulphuric acid rather than water. The presence of sulphate ions in the wash liquid ensures that the maximum amount of barium sulphate remains precipitated.

Did you know?

Kidney stones are produced by the precipitation of substances naturally present in the urine. The deposition depends on many factors. If the urine is too concentrated, the solubility product of some ions may be exceeded and precipitates such as calcium phosphate and calcium oxalate may build up. A higher concentration of calcium ions than normal may also shift the equilibrium in favour of the precipitation of insoluble calcium salts such as calcium carbonate and calcium phosphate, especially if the conditions become too alkaline.

Key points

- The common ion effect refers to the reduction in solubility of a dissolved salt by adding a solution which contains an ion in common with the dissolved salt.
- When the ionic product in the equilibrium expression for K_{sp} exceeds the solubility product a salt is precipitated.
- The solubility product of a salt can be found by determining the concentration of the ions in solution by titration or other methods.

9 Acid–base equilibria

9.1 Brønsted–Lowry theory of acids and bases

Learning outcomes

On completion of this section, you should be able to:

- explain the terms acid and base using Brønsted–Lowry theory
- describe the differences in behaviour of strong and weak acids and bases.

Simple definitions of acids and bases

An **acid** is a substance which neutralises a base to form a salt and water.

A **base** is a substance which neutralises an acid to form a salt and water:

$$\underset{\text{acid}}{2HCl(aq)} + \underset{\text{base}}{CaO(s)} \rightarrow \underset{\text{salt}}{CaCl_2(aq)} + \underset{\text{water}}{H_2O(l)}$$

Another definition of an acid is: a substance which ionises in water to form hydrogen ions. A hydrogen ion, H^+, is sometimes referred to as a proton.

$$HCl(g) + aq \rightarrow H^+(aq) + Cl^-(aq)$$

Another definition of a base is: a substance that dissolves or ionises in water to form hydroxide ions.

$$KOH(s) + aq \rightarrow K^+(aq) + OH^-(aq)$$

Many metal oxides or hydroxides are bases. A base which is soluble in water is called an **alkali**.

Did you know?

The H_3O^+ ion is called a **hydroxonium ion**. It is less frequently called a 'hydronium' or 'oxonium' ion. H^+ ions rarely occur alone in aqueous solutions. When we talk about hydrogen ions, we really mean a H_3O^+ ion. It is often convenient, however, especially in calculations to write this ion as H^+ and talk about hydrogen ions.

Brønsted–Lowry theory of acids and bases

- An acid is a proton donor.
- A base is a proton acceptor.

When hydrogen chloride gas reacts with water, the H^+ ion of the acid is donated to the water. So $HCl_{(g)}$ is a **Brønsted–Lowry acid**. The water accepts a proton so it is a Brønsted–Lowry base.

H^+ donated

$$\underset{\text{acid}}{HCl(g)} + \underset{\text{base}}{H_2O(l)} \rightarrow H_3O^+(aq) + Cl^-(aq)$$

When ammonia, NH_3, reacts with water, it accepts a proton from the water and becomes an NH_4^+ ion. So ammonia is a Brønsted–Lowry base. The water donates the proton, so it is a Brønsted–Lowry acid.

H^+ donated

$$\underset{\text{base}}{NH_3(g)} + \underset{\text{acid}}{H_2O(l)} \rightleftharpoons NH_4^+(aq) + OH^-(aq)$$

You will notice that water can act as an acid or a base – it is **amphoteric**.

Brønsted–Lowry acids do not have to involve aqueous solutions. For example, when methanoic acid reacts with chloric(III) acid, the following equilibrium is set up:

H^+ donated

$$\underset{\text{base}}{HCOOH} + \underset{\text{acid}}{HClO_2} \rightleftharpoons HCOOH_2^+ + ClO_2^-$$

Conjugate pairs

Every acid has a **conjugate base** – the ion left when an acid has given away its proton.

Every base has a **conjugate acid** – the ion formed when an acid has accepted a proton.

$$\underset{\text{acid}}{CH_3COOH\ (aq)} + \underset{\text{base}}{H_2O\ (l)} \rightleftharpoons \underset{\text{conjugate base}}{CH_3COO^-\ (aq)} + \underset{\text{conjugate acid}}{H_3O^+\ (aq)}$$

Strong and weak acids and bases

Strong acids ionise (almost) completely in solution. We sometimes say they dissociate completely. For example:

$$HNO_3(l) + H_2O(l) \rightarrow H_3O^+(aq) + NO_3^-(aq)$$

or more simply $HNO_3\ (l) + aq \rightarrow H^+(aq) + NO_3^-(aq)$

Hydrochloric acid, nitric acid and sulphuric acid are all strong acids.

Weak acids ionise (dissociate) only partially in solution. The position of equilibrium lies to the left. There are many more molecules of acid present than ions.

Organic acids such as methanoic acid, ethanoic acid and citric acid are weak acids.

$$CH_3COOH(l) + H_2O(l) \rightleftharpoons CH_3COO^-(aq) + H_3O^+(aq)$$

Strong bases ionise completely in solution, e.g.

$$NaOH(s) + aq \rightarrow Na^+(aq) + OH^-(aq)$$

Weak bases ionise (dissociate) only partially in solution. The position of equilibrium lies to the left. There are many more molecules of base present than ions.

Ammonia and amines are examples of weak bases.

$$CH_3NH_2(g) + H_2O(l) \rightleftharpoons CH_3NH_3^+(aq) + OH^-(aq)$$

Comparing weak acids and bases

The table shows some typical pH values for strong and weak acids and bases. Strong acids have much lower pH values than weak acids of the same concentration. Strong bases have much higher pH values than weak bases of the same concentration.

Concentration of acid or base	pH strong acid	pH weak acid	pH strong base	pH weak base
1.0 mol dm^{-3}	0	2.4	14	11.6
0.01 mol dm^{-3}	2	3.4	12	10.6

Strong acids have a higher electrical conductivity and react more rapidly with calcium carbonate or magnesium. This is because strong acids (for example hydrochloric acid) have a higher concentration of hydrogen ions than weak acids (for example ethanoic acid).

Exam tips

It is important to distinguish between the strength of an acid and its concentration. Strength refers to the degree of ionisation – strong or weak. Concentration refers to the number of moles per dm^3. A 4 mol dm^{-3} solution of ethanoic acid is concentrated, but it is a weak acid.

Key points

- The **Brønsted–Lowry theory** states that acids are proton donors and bases are proton acceptors.
- Strong acids and bases are (almost) completely ionised in aqueous solution. Weak acids and bases are only partially ionised in aqueous solution.
- Strong acids and bases of equal concentrations can be distinguished by the pH values of their aqueous solutions.

9.2 Simple pH calculations

Learning outcomes

On completion of this section, you should be able to:

- define the terms pH and K_w
- perform calculations involving pH, hydrogen and hydroxide ion concentration and K_w.

pH

We can compare the relative acidity or alkalinity of substances using the **pH** scale. We express this as the logarithm to the base 10 of the hydrogen ion concentration (in mol dm^{-3}):

$$pH = -\log_{10}[H^+]$$

Most pH values we use in chemistry are on a scale of 0 to 14 with neutrality (neither acidic nor alkaline) being pH 7. Very concentrated acids can have pH values below 0 and very concentrated alkalis can have pH values above 14.

The p notation

The p notation (as in pH) is a modified logarithmic scale used by chemists to express small values simply. The relationship between the numbers and $-\log_{10}$ is shown below.

Number	1×10^{-12}	1×10^{-7}	1×10^{-5}	10^0	1×10^4
$-\log$ (number)	12	7	5	0	-4

The p notation can also be used in relation to other equilibrium constants, e.g.

$$pOH = -\log_{10}[OH^-] \quad pK_a = -\log_{10}[K_a] \quad pK_w = -\log_{10}[K_w]$$

Simple pH calculations

The pH of strong acids

Strong acids are completely ionised in water. So the pH can be easily calculated.

$$1\,\text{mol dm}^{-3}\,\text{HCl} = -\log(1 \times 10^{-1}) = \text{pH } 0$$
$$0.01\,\text{mol dm}^{-3}\,\text{HCl} = -\log(1 \times 10^{-2}) = \text{pH } 2$$
$$0.002\,\text{mol dm}^{-3}\,\text{HCl} = -\log(2 \times 10^{-3}) = \text{pH } 2.7$$

Exam tips

You must make sure that you know how to convert pH to H^+ concentration and H^+ concentration to pH. This may vary from calculator to calculator. The most common method is:

pH e.g. pH 3.6 to $[H^+]$

3.6 → ± → (shift) 10^x → answer (2.51×10^{-4})

$[H^+]$ e.g. 1.8×10^{-5} to pH

1.8 → EXP → 5 → ± → log → ± → answer (4.74)

If you are not sure – ask your teacher.

Worked example 1

Calculate the pH of a solution whose hydrogen ion concentration is $3.32 \times 10^{-5}\,\text{mol dm}^{-3}$

$$\begin{aligned} pH &= -\log_{10}[H^+] \\ &= -\log_{10}(3.32 \times 10^{-5}) = 4.48 \text{ (note that there are no units to pH)} \end{aligned}$$

Worked example 2

Calculate the hydrogen ion concentration of a solution of pH 10.5.

$$\begin{aligned} pH &= -\log_{10}[H^+] \\ [H^+] &= 10^{-pH} = 10^{-10.5} = 3.2 \times 10^{-10}\,\text{mol dm}^{-3} \end{aligned}$$

The ionic product of water, K_w

Water can act as either an acid or base depending on what type of substance is dissolved in it (see Section 9.1). In pure water, a few water molecules can act as an acid by donating protons to other water molecules.

H_2O	+	H_2O	⇌	H_3O^+	+	OH^-
acid		base		conjugate acid		conjugate base

We can simplify this equation to: $H_2O \rightleftharpoons H^+ + OH^-$

For which the equilibrium expression is: $K = \dfrac{[H^+][OH^-]}{[H_2O]}$

Since the concentration of water is very high, we can incorporate this into the value for K. The expression then becomes:

$$K_w = [H^+][OH^-]$$

K_w is called the **ionic product of water**. Its value is $1.00 \times 10^{-14}\,\text{mol}^2\text{dm}^{-6}$.

Using K_w to calculate the pH of alkalis

We can use K_w to calculate the pH of strong bases. We need to know the concentration of hydroxide ions in the solution and value of K_w.

Worked example 3

Calculate pH of a solution of potassium hydroxide of concentration $0.0250\,\text{mol dm}^{-3}$.

Step 1: Write the ionic product expression in terms of $[H^+]$ ions.

$$K_w = 1.00 \times 10^{-14}\,\text{mol}^2\text{dm}^{-6}$$

Step 2: Write the equilibrium expression in terms of $[H^+]$

$$[H^+] = \frac{K_w}{[OH^-]}$$

Step 3: Substitute the values into the expression.

$$[H^+] = \frac{1.00 \times 10^{-14}}{0.0250} = 4.00 \times 10^{-13}$$

Step 4: Calculate pH: $-\log_{10}(4.00 \times 10^{-13}) = \text{pH } 12.4$

Did you know?

Why is the pH of water 7?

We can see this if we use the ionic expression for water:

$$[H^+][OH^-] = 1.00 \times 10^{-14}$$

and $[H^+] = [OH^-]$

So $[H^+]^2 = 1.00 \times 10^{-14}$

and $[H^+] = \sqrt{1.00 \times 10^{-14}} = 1.00 \times 10^{-7}$

$-\log_{10}(1.00 \times 10^{-7}) = 7$

Key points

- pH is a measure of the hydrogen ion concentration, $[H^+]$.
- $pH = -\log_{10}[H^+]$.
- K_w is the ionic product of water (value $= 1.00 \times 10^{-14}\,\text{mol}^2\text{dm}^{-6}$).
- The pH of a strong base can be found using the expression $K_w = [H^+][OH^-]$.

9.3 Equilibrium calculations involving weak acids

Learning outcomes

On completion of this section, you should be able to:

- understand the term 'acid dissociation constant, K_a'
- perform calculations involving the use of K_a, pH and $[H^+]$.

The acid dissociation constant, K_a

The equilibrium law can be applied to aqueous solutions of weak acids. The equation for the partial ionisation of ethanoic acid in water is:

$$CH_3COOH(aq) + H_2O(l) \rightleftharpoons CH_3COO^-(aq) + H_3O^+(aq)$$

Because water is present at a very high concentration, we can assume that its concentration is constant and we can simplify this equation to:

$$CH_3COOH(aq) \rightleftharpoons CH_3COO^-(aq) + H^+(aq)$$

The equilibrium expression for this reaction is:

$$K_a = \frac{[CH_3COO^-][H^+]}{[CH_3COOH]}$$

- K_a is the **acid dissociation constant**. For a monobasic weak acid such as ethanoic acid, the units of K_a are $mol\,dm^{-3}$.
- A high value for K_a e.g. $40\,mol\,dm^{-3}$ indicates that the position of equilibrium is well over to the right. The acid is almost completely ionised.
- A low value for K_a e.g. $1.3 \times 10^{-5}\,mol\,dm^{-3}$ indicates that the position of equilibrium is well over to the left. The acid is only very slightly ionised.
- We can also use pK_a values to compare the strengths of acids. $pK_a = -\log_{10}[K_a]$.

The general equilibrium expression for K_a

- We can write a general equilibrium expression for all monobasic acid based on the general reaction:

$$HA \rightleftharpoons H^+ + A^-$$

- HA represents the unionised acid.
- A^- represents the conjugate base of the acid.

The general equilibrium expression is therefore:

$$K_a = \frac{[H^+][A^-]}{[HA]}$$

Since the concentration of H^+ and A^- must be equal, we can simplify this further:

$$K_a = \frac{[H^+]^2}{[HA]}$$

In calculations involving K_a we make two assumptions.

- If the K_a value for the weak acid is very small, we can assume that the concentration of acid molecules remaining at equilibrium is the same as the concentration of undissociated acid molecules in the original acid.
- We can ignore the hydrogen ions which arise from the ionisation of water. For most work this is acceptable because the ionic product for water, K_w, is very small.

Calculations involving K_a

We can calculate the value of K_a for a weak acid if we know:

- the concentration of acid
- the pH of the solution.

We can calculate the pH of a solution of a weak acid if we know

- the value of K_a
- the concentration of the acid.

Worked example 1

Deduce the pH of a solution of 0.05 mol dm^{-3} ethanoic acid.

K_a for ethanoic acid = 1.74×10^{-5} mol dm^{-3}.

Step 1: Write the equilibrium expression.

$$K_a = \frac{[H^+]^2}{[CH_3COOH]} \text{ or } K_a = \frac{[H^+]^2}{[HA]}$$

Step 2: Rearrange the equation to make $[H^+]^2$ the subject:

$$[H^+]^2 = K_a \times [CH_3COOH]$$

Step 3: Substitute the values: $[H^+]^2 = (1.74 \times 10^{-5}) \times 0.05 = 8.7 \times 10^{-7}$

Step 4: Find $[H^+]$ by taking the square root:
$\sqrt{8.7 \times 10^{-7}} = 9.33 \times 10^{-4}$ mol dm^{-3}

Step 5: Convert $[H^+]$ to pH: pH = $-\log_{10}[H^+] = -\log_{10}(9.33 \times 10^{-4})$
pH = 3.03

Worked example 2

The pH of a 0.01 mol dm^{-3} of methanoic acid is 2.9. Deduce the value of K_a.

Stage 1: Convert pH to $[H^+]$: $[H^+] = 10^{-2.9} = 1.26 \times 10^{-3}$ mol dm^{-3}

Stage 2: Write the equilibrium expression.

$$K_a = \frac{[H^+]^2}{[HCOOH]} \text{ or } K_a = \frac{[H^+]^2}{[HA]}$$

Stage 3: Substitute into the equilibrium expression:

$$K_a = \frac{(1.26 \times 10^{-3})^2}{0.01}$$

$$K_a = 1.59 \times 10^{-4} \text{ mol dm}^{-3}$$

Key points

- For a weak acid, $[H^+]$ and pH can be calculated using the expression:

$$K_a = \frac{[H^+][A^-]}{[HA]}$$

Where $[A^-]$ is the concentration of ionised acid and $[HA]$ is the concentration of unionised acid.

Exam tips

In calculations involving K_a, remember that hydrogen ion concentration can be converted to pH and pH to hydrogen ion concentration.

9.4 Acid and base dissociation constants

Learning outcomes

On completion of this section, you should be able to:

- describe pK_a, K_b and pK_b
- perform calculations involving pK_a, K_b and pK_b and pK_w.

Exam tips

We can calculate the approximate pH for a dibasic acid such as H_2S in a similar way as for a monobasic acid. This is because the value of K_{a1} is much greater than K_{a2}.

e.g.

$$H_2S(aq) \rightleftharpoons HS^-(aq) + H^+(aq)$$
$$K_{a1} = 8.9 \times 10^{-8}\ mol\ dm^{-3}$$
$$HS^-(aq) \rightleftharpoons S^{2-}aq) + H^+(aq)$$
$$K_{a2} = 1.2 \times 10^{-13}\ mol\ dm^{-3}$$

We can ignore the H^+ ions arising from the second ionisation because their concentration is insignificant compared with those arising form the first ionisation.

Using pK_a values

We can use pK_a values to compare the acidity of weak acids. $pK_a = -\log_{10}[K_a]$. Some examples are shown in the table.

Acid	Equilibrium in aqueous solution	K_a/mol dm^{-3}	pK_a
sulphuric(IV)	$H_2SO_3 \rightleftharpoons H^+ + HSO_3^-$	1.5×10^{-2}	1.82
hydrofluoric	$HF \rightleftharpoons H^+ + F^-$	5.6×10^{-4}	3.25
methanoic acid	$HCOOH \rightleftharpoons H^+ + HCOO^-$	1.6×10^{-4}	3.80
'carbonic acid'	$CO_2 + H_2O \rightleftharpoons H^+ + HCO_3^-$	4.5×10^{-7}	6.35

The higher the pK_a value, the weaker the acid.

We can use pK_a values in calculations rather than K_a values.

Worked example 1

Deduce the pH of a solution of 0.001 mol dm^{-3} benzoic acid. pK_a benzoic acid = 4.2.

Step 1: Convert pK_a to K_a: $pK_a = -\log_{10}[K_a]$
So $K_a = 10^{-pK_a}$ $K_a = 10^{-4.2}$ $K_a = 6.3 \times 10^{-5}$ mol dm^{-3}

Step 2: Write the equilibrium expression:

$$K_a = \frac{[H^+]^2}{[HA]}$$

Step 3: Rearrange the equation to make $[H^+]^2$ the subject:

$$[H^+]^2 = K_a \times [HA]$$

Step 4: Substitute the values: $[H^+]^2 = (6.3 \times 10^{-5}) \times 0.001 = 6.3 \times 10^{-8}$

Step 5: Find $[H^+]$ by taking the square root:

$$\sqrt{6.3 \times 10^{-8}} = 2.51 \times 10^{-4}\ mol\ dm^{-3}$$

Step 6: Convert $[H^+]$ to pH: $pH = -\log_{10}[H^+] = -\log_{10}(2.51 \times 10^{-4})$
pH = 3.6

The basic dissociation constant, K_b

For calculations involving a weak alkali, we can use the basic dissociation constant, K_b. For example, in the reaction:

$$NH_3(g) + H_2O(l) \rightleftharpoons NH_4^+(aq) + OH^-(aq)$$

We can write the equilibrium expression:

$$K_b = \frac{[NH_4^+][OH^-]}{[NH_3]} \quad \text{or more generally} \quad K_b = \frac{[BH^+][OH^-]}{[B]}$$

Where [B] is the weak base and $[BH^+]$ is the conjugate acid.

We make similar assumptions to those for a weak acid in applying this expression (see Section 9.1).

Since $[BH^+] = [OH^-]$ we can simplify the expression further to:

$$K_b = \frac{[OH^-]^2}{[B]}$$

Worked example 2

Calculate the pH of a 0.01 mol dm^{-3} solution of ethylamine.

K_b (ethylamine) = 6.5×10^{-4} mol dm^{-3} $K_w = 1.0 \times 10^{-14}$ mol dm^{-3}

Step 1: Write the equilibrium expression:

$$K_b = \frac{[OH^-]^2}{[B]}$$

Step 2: Rearrange the equation to make $[OH^-]^2$ the subject:

$$[OH^-]^2 = K_b \times [B]$$

Step 4: Substitute the values: $[OH^-]^2 = (6.5 \times 10^{-4}) \times 0.01 = 6.5 \times 10^{-6}$

Step 5: Find $[OH^-]$ by taking the square root:

$$\sqrt{6.5 \times 10^{-6}} = 2.55 \times 10^{-3}\,\text{mol dm}^{-3}$$

Step 6: Use $K_w = [H^+][OH^-]$ to calculate $[H^+]$:

$$[H^+] = \frac{K_w}{[OH^-]} = \frac{1.0 \times 10^{-14}}{2.55 \times 10^{-3}} = 3.92 \times 10^{-12}\,\text{mol dm}^{-3}$$

Step 7: Calculate pH: $\text{pH} = -\log_{10}[H^+] = -\log_{10}(3.92 \times 10^{-12})$
$\text{pH} = 11.4$

Using pK_a and pK_b values

A useful expression which relates K_w to K_a and K_b is

$$pK_a + pK_b = pK_w$$

This can also be written as:

$$-\log_{10}K_a - \log_{10}K_b = 14 \text{ (the 14 derives from } -\log_{10}(1.0 \times 10^{-14}))$$

Exam tips

A quick way to get an answer to a question such as 'calculate the pH of a solution of a strong base of concentration 0.1 mol dm^{-3}' is to use the expression $14 - \log_{10}[OH^-]$. This works because $-\log_{10}[H^+] - \log_{10}[OH^-] = 14$

Worked example 3

Deduce the value of K_a for a 0.01 mol dm^{-3} solution of aqueous ammonia.

pK_b for ammonia = 4.7

Step 1: Apply the relationship $pK_a + pK_b = pK_w$ to calculate pK_a:

$$pK_a + 4.7 = 14 \quad \text{So } pK_a = 14 - 4.7 = 9.3$$

Step 2: Convert pK_a to K_a:

$$pK_a = -\log_{10}K_a = -\log_{10}(9.3) \quad K_a = 5.0 \times 10^{-10}\,\text{mol dm}^{-3}$$

Key points

- For a weak base the concentration of hydroxide ions can be calculated using the general expression:

$$K_b = \frac{[BH^+][OH^-]}{[B]}$$

  where $[BH^+]$ is the concentration of the protonated base (salt) and [B] is the concentration of unionised base.
- $pK_a = -\log K_a$ and $pK_b = -\log K_b$

9.5 Changes in pH during acid–base titrations

Learning outcomes

On completion of this section, you should be able to:

- describe changes in pH during acid–base titrations
- explain the changes in pH when a strong or weak acid reacts with a strong or weak base.

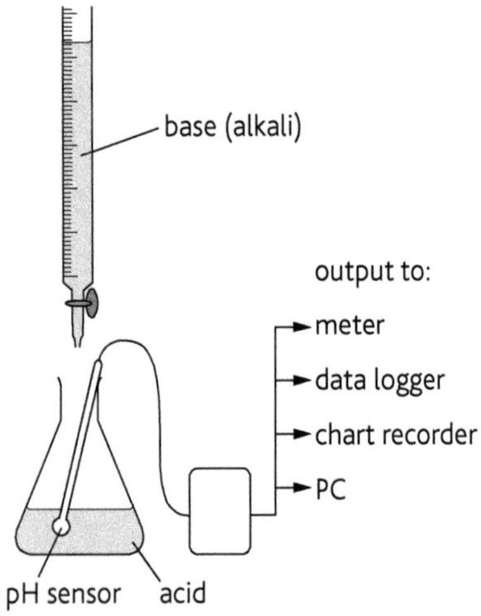

Figure 9.5.1 *Monitoring the pH as a base is added to an acid*

pH–titration curves

Measuring pH changes

Titration curves are graphs showing how the pH of an acid or base changes when an acid is added to an alkali or an alkali added to an acid. Figure 9.5.1 shows the apparatus used to follow these changes. A pH electrode is placed in the acid in the flask and the alkali added gradually from the burette at a slow and constant rate. The solution is kept stirred all the time. The pH is recorded either:

- continuously using a data logger attached to a computer or
- manually by recording the pH after fixed volumes of acid have been added to the alkali in the flask.

Titration curves

The graphs in Figure 9.5.2 show the results obtained when strong and weak monoprotic acids are titrated with alkalis. Monoprotic acids are acids in which only one hydrogen ion per molecule of acid can be donated when it reacts. The exact shape of the curve depends on whether the acid and base are strong or weak.

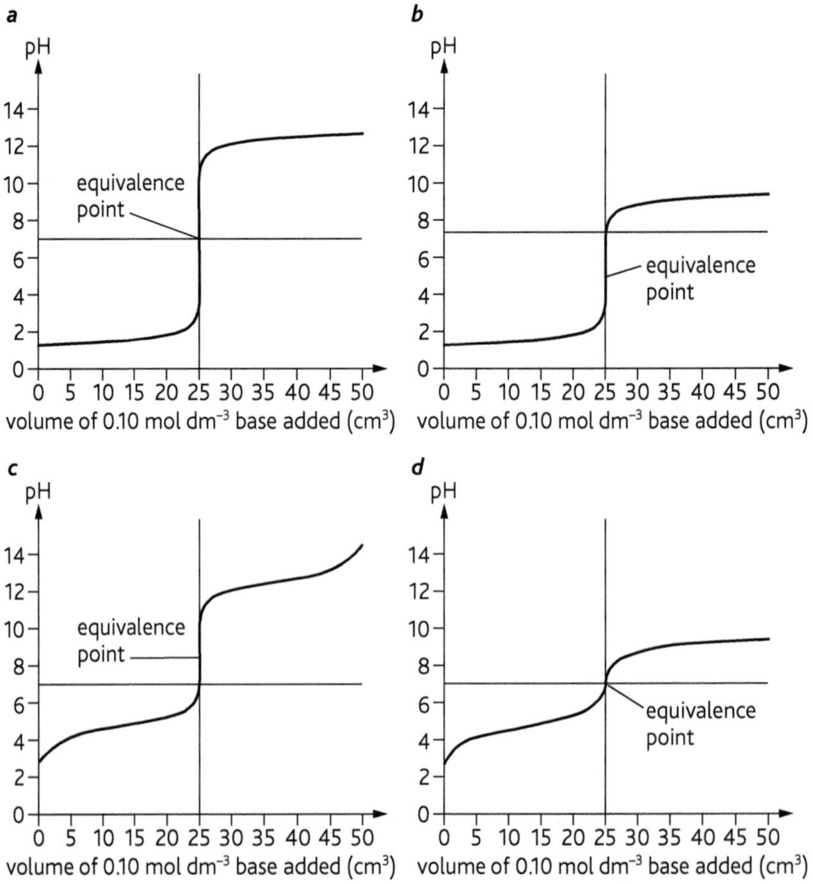

Figure 9.5.2 *pH titration curves for:* ***a*** *a strong acid and a strong base;* ***b*** *a strong acid and a weak base;* ***c*** *a weak acid and a strong base;* ***d*** *a weak acid and a weak base*

Apart from the curve for the weak acid and weak base, the curves show a very steep portion where a single drop of base changes the pH by several units. The point at which the base has exactly neutralised the acid is called the **equivalence point**. You can see that this corresponds to the steep portion of the curve.

Strong acid and strong base

Figure 9.5.2(a) shows the titration of 0.10 mol dm^{-3} hydrochloric acid (strong acid) with 0.10 mol dm^{-3} sodium hydroxide (strong base).

$$HCl(aq) + NaOH(aq) \rightarrow NaCl(aq) + H_2O(l)$$

We can see why the pH changes so rapidly around the equivalence point by calculation. Just before the equivalence point there is 24.9 cm^3 of NaOH and 25 cm^3 HCl present. The excess 0.1 cm^3 of acid is present in 49.9 cm^3 solution.

So $[H^+] = 0.1 \times \frac{0.1}{49.9} = 0.0002\,mol\,dm^{-3}$ = pH 3.7

When there is 0.1 cm^3 of alkali in excess: $[OH^-] = 0.0002\,mol\,dm^{-3}$ $[H^+] = 1.0 \times 10^{-14}/0.0002 = 5 \times 10^{-11}\,mol\,dm^{-3}$ So pH = 10.3.

The equivalence point is at approximately pH 7 because the salt formed is neutral.

Strong acid and weak base

Figure 9.5.2(b) shows the titration of 0.10 mol dm^{-3} hydrochloric acid (strong acid) with 0.10 mol dm^{-3} aqueous ammonia (weak base).

$$HCl(aq) + NH_3(aq) \rightarrow NH_4Cl(aq)$$

The ammonium chloride formed is the salt of a strong acid and a weak base and is acidic as a result of hydrolysis:

$$NH_4Cl(aq) + H_2O(l) \rightarrow NH_3(aq) + H_3O^+(aq) + Cl^-(aq)$$

So the equivalence point is at about pH 5.

Strong base and weak acid

Figure 9.5.2(c) shows the titration of 0.10 mol dm^{-3} ethanoic acid (weak acid) with 0.10 mol dm^{-3} sodium hydroxide (strong base).

$$CH_3COOH(aq) + NaOH(aq) \rightarrow CH_3COO^-Na^+(aq) + H_2O(l)$$

The sodium ethanoate formed is the salt of a weak acid and a strong base and is alkaline as a result of hydrolysis:

$$CH_3COO^-(aq) + H_2O(l) \rightarrow CH_3COOH(aq) + OH^-(aq)$$

So the equivalence point is at about pH 9.

Weak base and weak acid

Figure 9.5.2(d) shows the titration of 0.10 mol dm^{-3} ethanoic acid (weak acid) with 0.10 mol dm^{-3} aqueous ammonia (weak base).

$$CH_3COOH(aq) + NH_3(aq) \rightarrow CH_3COO^-NH_4^+(aq)$$

The ammonium ethanoate formed is the salt of a weak acid and a weak base. Both the ethanoate and ammonium ions can undergo hydrolysis leaving the equivalence point at around pH 7.

$$CH_3COO^-(aq) + H_2O(l) \rightarrow CH_3COOH(aq) + OH^-(aq)$$

Exam tips

If you titrate a base (in the flask) with an acid (in the burette), the pH titration curves will be the mirror images of those that are shown, e.g. for a strong acid and a strong base:

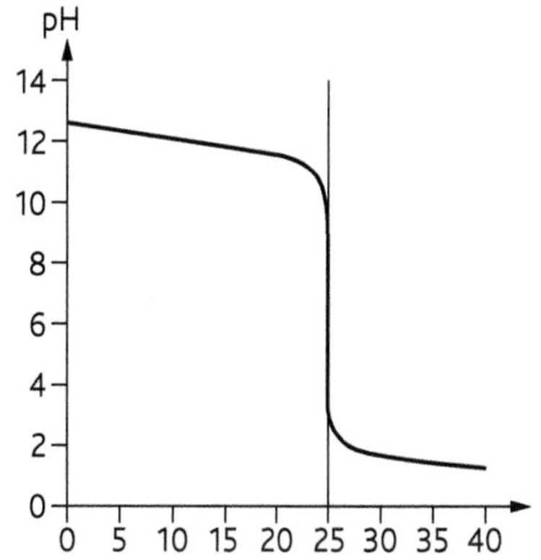

Key points

- The shape of a pH–titration curve depends on the strength of the acid and base used.
- For a strong acid and a strong base there is a very large and sudden pH change corresponding to the end point of the titration.
- For a strong acid and a weak base or a weak acid and a strong base, there is a fairly large and sudden pH change corresponding to the end point of the titration.
- For a weak acid and a weak base there is a gradual change in pH as the acid is added to the base.

9.6 Acid–base indicators

Learning outcomes

On completion of this section, you should be able to:

- know that specific indicators are used to determine the end point of acid–base titrations depending on the strength of the acid and base
- explain the term 'pH range' as applied to indicators
- explain the basis for selecting appropriate indicators for use in acid–base titrations.

Introduction to acid–base indicators

An **acid–base indicator** is a substance which changes colour when the pH changes over a certain range of pH values. Indicators can be thought of as being weak acids whose conjugate base has a different colour.

$$\underset{\substack{\text{indicator molecule} \\ \text{colour A}}}{\text{HIn}} \rightleftharpoons \text{H}^+ + \underset{\substack{\text{deprotonated indicator} \\ \text{colour B}}}{\text{In}^-}$$

- In more acidic (or less alkaline) conditions the concentration of H^+ ions is higher.
- So the position of equilibrium moves to the left.
- So the indicator shows colour A.
- In more alkaline (or less acidic) conditions the concentration of H^+ ions is lower.
- So the position of equilibrium moves to the right.
- So the indicator shows colour B.

Indicator range

Indicators usually change colour over a **pH range** of one or two units. For indicators to be useful, the colour of the indicator should change in the middle of its range. At this point, the colour of the indicator may be intermediate between the two extremes. For example, bromothymol blue is yellow at pH 6 and blue at pH 7.6. At an intermediate point it appears a greyish green.

$$\begin{array}{ccccc} \text{pH 6} & \longleftarrow & \text{adding acid to alkali} & \text{——} & \text{pH 7.6} \\ \uparrow & & \uparrow & & \uparrow \\ \text{yellow} & & \text{grey-green} & & \text{blue} \end{array}$$

The colour change of an indicator should also be distinct. Methyl orange is not a good indicator alone since the colour change is rather indistinct. That is why screened methyl orange (a mixed indicator) is often used in preference.

For an indicator to work correctly, the colour must change sharply at the end point of the titration – the vertical part of the curve. For this to happen, the rapid change in pH in the region of the end point must correspond to working range of the indicator.

The table below shows some colour changes and the working range of some indicators.

Indicator	Colour at lower pH in the range	Colour at higher pH in the range	pH at end point
methyl orange	red at pH 3.2	yellow at pH 4.4	3.7
methyl red	red at pH 4.2	yellow at pH 6.3	5.1
bromothymol blue	yellow at pH 6.0	blue at pH 7.6	7.0
phenolphthalein	colourless at pH 8.2	deep pink at pH 10	9.3
alizarin yellow	yellow at pH 10.1	reddish at pH 13.0	12.5

Choosing the correct indicator

The selection of a correct indicator depends on where the most rapid change in pH occurs as the acid is titrated with a base.

Strong acid and strong base

Most indicators can be used for a strong acid and a strong base, e.g. methyl orange, phenolphthalein, bromothymol blue, because their range falls within the vertical part of the curve. Phenolphthalein is often preferred because its colour change is more obvious.

Strong acid and weak base

The sharpest change in the pH curve for a strong acid and a weak base is between pH 3.0 and 7.5. Methyl red, methyl orange or bromothymol blue are suitable indicators because their range falls within the vertical part of the curve. Phenolphthalein is unsuitable because its colour change does not correspond to the vertical part of the curve. It would only change colour slowly after the equivalence point.

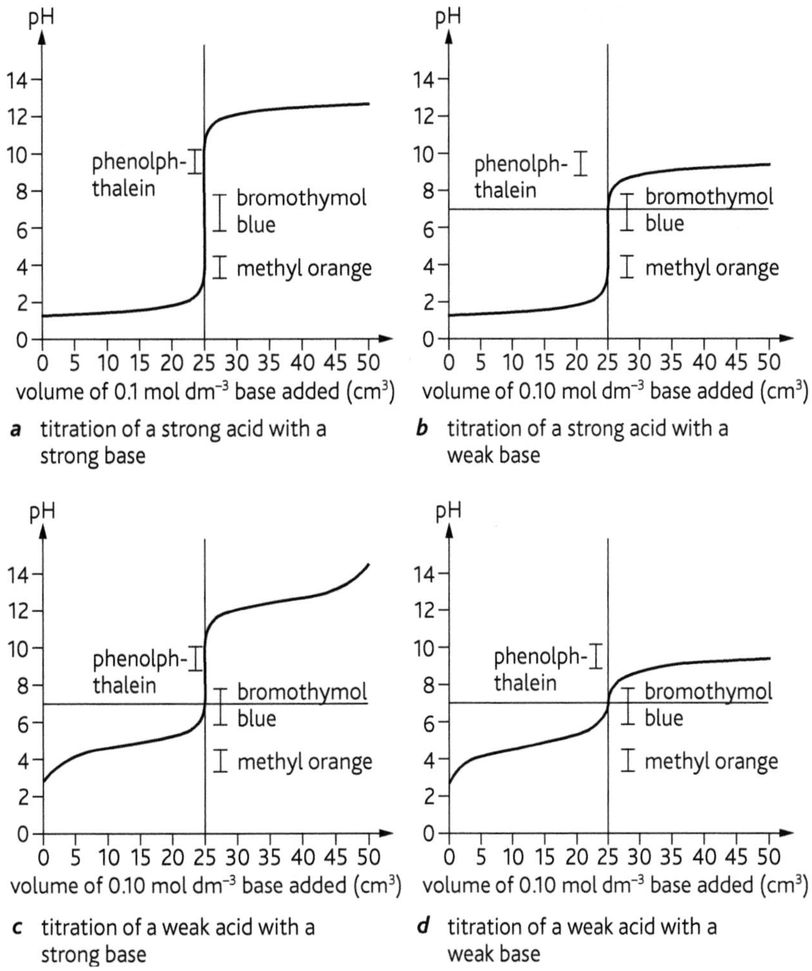

***Figure 9.6.1** Selecting a suitable indicator for different acid–base titrations: **a** a strong acid and a strong base; **b** a strong acid and a weak base; **c** a weak acid and a strong base; **d** a weak acid and a weak base*

Weak acid and strong base

The sharpest change in the pH curve for a weak acid and a strong base is between pH 7.0 and 11. Phenolphthalein is a suitable indicator because its colour change corresponds to the vertical part of the curve. Methyl red and methyl orange are not suitable indicators because their range does not fall within the vertical part of the curve. They would only change colour slowly before the equivalence point.

Weak acid and weak base

No indicator is suitable for a weak acid and a weak base. None of the indicators would show a sudden change in colour because the pH is changing slowly. For example, bromothymol blue starts changing colour when about 24 cm^3 of alkali have been added and finishes changing colour when 26 cm^3 have been added.

Key points

- Acid–base indicators change colour when there is a sharp change in pH.
- The pH range of an indicator is the range (usually about 2 pH units) over which the indicator changes colour.
- For an indicator to work correctly in an acid–base titration, the colour change must correspond to the place in the pH-titration curve where pH changes suddenly.

9.7 Buffer solutions

Learning outcomes

On completion of this section, you should be able to:

- define the term 'buffer solution'
- explain how buffer solutions control pH
- describe the importance of buffer solutions in biological systems and industrial processes.

What is a buffer solution?

A **buffer solution** is a solution which minimises pH changes when small amounts of acids or alkalis are added.

Acidic buffers

An acidic buffer solution is a mixture of a weak acid and its conjugate base (the salt of the weak acid). An example is an aqueous solution of ethanoic acid (0.1 mol dm^{-3}) and its conjugate base, sodium ethanoate (0.1 mol dm^{-3}).

$$\underset{\text{weak acid}}{CH_3COOH(aq)} \rightleftharpoons \underset{\text{conjugate base (salt)}}{CH_3COO^-(aq)} + H^+(aq)$$

So a buffer solution contains relatively high concentrations of both acid and conjugate base. We say that there are reserve supplies of base.

If the ratio of concentration of conjugate base (added salt) and acid does not change very much, the hydrogen ion concentration (and therefore the pH) will not change very much (see also Section 9.3).

On addition of acid to this buffer solution:

- H^+ ions combine with CH_3COO^- ions.
- So the position of equilibrium shifts to the left and a little more undissociated acid is formed.
- But because of the relatively high concentration of added base CH_3COO^-, the ratio of base to acid changes very little.
- So the pH hardly changes.

On addition of alkali to this buffer solution:

- OH^- ions combine with H^+ ions to form water.
- The removal of H^+ ions shifts the position of equilibrium to the right and a little more acid, CH_3COOH dissociates to CH_3COO^-.
- But because of the relatively high concentration of added base CH_3COO^-, the ratio of base to acid changes very little.
- So the pH hardly changes.

A buffer may be made more effective by increasing the concentration of both the acid and the conjugate base (salt).

If the ratio of concentration of conjugate base (added salt) and acid does not change very much, the hydrogen ion concentration (and therefore the pH) will not change very much. If, however, very large amounts of acid or alkali are added, the buffer will not work because the position of equilibrium has been so far altered as to lower the concentration of either the acid or its conjugate base by an unacceptable amount.

Exam tips

Remember that most of the conjugate base in acidic buffer solutions is a salt. It is completely ionised. You may see salt and conjugate base used interchangeably.

Basic buffers

A basic buffer solution is a mixture of a weak base and its conjugate acid (salt). An example is an aqueous solution of ammonia (0.1 mol dm^{-3}) and its conjugate base, ammonium chloride (0.1 mol dm^{-3}).

$$\underset{\text{weak base}}{NH_3(aq)} + H^+(aq) \rightleftharpoons \underset{\text{conjugate acid}}{NH_4^+(aq)}$$

This buffer solution contains relatively high concentrations of both base and conjugate acid. There are reserve supplies of acid.

On addition of acid to this buffer solution, the aqueous ammonia removes the added H^+ ions. A little more salt, NH_4^+, is formed.

On addition of alkali to this buffer solution, the aqueous ammonia removes the added H^+ ions to form water. The position of equilibrium shifts to the left and a little more ammonia is formed.

Buffers in natural systems

Many enzymes will only work at specific pH values near pH 7. So animals depend on buffer systems to keep the pH constant in various parts of their bodies. The pH of the blood is approximately pH 7.4. A number of blood buffers help to maintain this pH.

Hydrogencarbonate buffer: Carbon dioxide (a product of respiration) combines with water in the blood to form a solution of hydrogencarbonate ions.

$$CO_2(g) + H_2O(l) \rightleftharpoons HCO_3^-(aq) + H^+(aq)$$

If the H^+ ion concentration in the blood rises:

- H^+ ions combine with HCO_3^- ions.
- The position of equilibrium moves slightly to the left.
- The concentration of H^+ ions is reduced to keep the pH constant.
- There is still a high enough concentration of HCO_3^- ions to allow the buffer to function.

If the OH^- ion concentration in the blood rises, they combine with H^+ ions to form water. The equilibrium moves slightly to the right.

Phosphate buffers: Several type of phosphate ions are present in the blood plasma. The equilibrium involved is:

$$\underset{\text{acid}}{H_2PO_4^-} \rightleftharpoons \underset{\text{conjugate base}}{HPO_4^{2-}} + H^+$$

If blood gets too acid, the position of equilibrium shifts slightly to the left but there are still sufficient $H_2PO_4^-$ and HPO_4^{2-} in solution for the phosphates to act as a buffer.

Protein buffers: Some proteins and amino acids in the blood act as buffers. These buffers can be acidic or basic depending on the charges on the amino acids or proteins:

$$\underset{\text{protonated form}}{PH} \rightleftharpoons \underset{\text{deprotonated form}}{P^-} + H^+ \quad \text{or} \quad \underset{\text{protonated form}}{PH^+} \rightleftharpoons \underset{\text{deprotonated form}}{P} + H^+$$

Buffer solutions in industry

Buffer solutions are used in electroplating, the manufacture of dyes and treatment of leather. They are also present in some detergents, soaps and shampoos where it is important not to damage fabric, skin or hair.

Key points

- A buffer solution minimises pH change on addition of a small amount of acid or base.
- A buffer solution is a mixture of a weak acid and its conjugate base or a weak base and its conjugate acid.
- Buffer solutions maintain pH by keeping a fairly constant ratio of acid and conjugate base.
- Blood buffer solutions include HCO_3^- ions, phosphate ions and proteins.

9.8 Calculations involving buffer solutions

Learning outcomes

On completion of this section, you should be able to:

- know how to calculate the pH of a buffer solution from appropriate data
- know how to calculate the acid/base ratio required to make a buffer solution with a particular pH
- know how to conduct an experiment to determine the pH of a buffer solution.

Equilibrium aspects

Consider a buffer solution consisting of a weak acid and its conjugate base:

$$\underset{\text{weak acid}}{CH_3COOH(aq)} \rightleftharpoons \underset{\text{conjugate base}}{CH_3COO^-(aq)} + H^+(aq)$$

We can write an equilibrium expression for this reaction in terms of hydrogen ions:

$$K_a = \frac{[CH_3COO^-][H^+]}{[CH_3COOH]}$$

So $[H^+] = K_a \times \frac{[CH_3COOH]}{[CH_3COO^-]}$ or more generally $[H^+] = K_a \times \frac{[HA]}{[A^-]}$

Since K_a is constant, the ratio of the concentration of acid to conjugate base (salt) controls the H^+ ion concentration (and thus the pH). Dilution of the buffer has no effect on its pH since A^- and HA are diluted equally.

Deducing the pH of a buffer solution

We can calculate the pH of a buffer solution if we know:

- the value of K_a of the weak acid
- the equilibrium concentrations of the weak acid and conjugate base.

Worked example 1

Deduce the pH of a buffer solution made by adding 0.20 mol of sodium ethanoate to 500 cm^3 of 0.10 mol dm^{-3} ethanoic acid. K_a of ethanoic acid = 1.7×10^{-5} mol dm^{-3}.

Step 1: Calculate the concentrations of the acid and its conjugate base:

$$[CH_3COOH] = 0.10\,\text{mol dm}^{-3}$$

$$[CH_3COO^-] = 0.2 \times \frac{1000}{500} = 0.40\,\text{mol dm}^{-3}$$

Step 2: Rearrange the equilibrium expression in terms of $[H^+]$ ions:

$$[H^+] = K_a \times \frac{[CH_3COOH]}{[CH_3COO^-]}$$

Step 3: Substitute the values:

$$[H^+] = 1.7 \times 10^{-5} \times \frac{0.10}{0.40} = 4.25 \times 10^{-6}\,\text{mol dm}^{-3}$$

Step 4: Calculate pH: pH = $-\log_{10}[H^+] = -\log_{10}(4.25 \times 10^{-6})$ pH = 5.4 (to 2 s.f.)

We can make buffer solution calculations easier to deal with if we use the expression:

$$pH = pK_a + \log_{10}\frac{[\text{salt}]}{[\text{acid}]}$$

So if we are given the pK_a or calculate it from K_a, we can do the calculation much more simply.

Step 1: Convert K_a to pK_a: $-\log_{10}(1.7 \times 10^{-5})$ $\quad pK_a = 4.77$

Step 2: Substitute into the equation:

$$pH = 4.77 + \frac{\log_{10}(0.40)}{(0.10)}$$

$$pH = 4.77 + 0.6$$
$$pH = 5.4 \text{ (to 2 s.f.)}$$

Calculating the acid–conjugate base ratio

For an acidic buffer solution, the higher the acid–conjugate base ratio, the lower is the pH of the buffer. But in order to make a buffer solution with a particular pH, we need to know exactly how much base to add to an acid or in what volumes to mix solutions of the acid and salt of known concentrations. The example below shows how to do this. Remember that [salt] is the same as [conjugate base].

Worked example 2

How many moles of sodium propanoate must be added to 250 cm^3 of solution containing 0.10 mol of propanoic acid to make a buffer solution of pH 5.30?

K_a of propanoic acid = $1.35 \times 10^{-5}\,mol\,dm^{-3}$.

Step 1: Calculate the concentration of the propanoic acid:

$$[acid] = 0.10 \times \frac{1000}{250} = 0.40\,mol\,dm^{-3}$$

Step 2: Calculate $[H^+]$ from pH: $[H^+] = 10^{-5.30} = 5.01 \times 10^{-5}\,mol\,dm^{-3}$

Step 3: Rearrange the equilibrium expression with [salt] as the subject:

$$[salt] = K_a \times \frac{[acid]}{[H^+]}$$

Step 4: Substitute the values:

$$[salt] = \frac{(1.35 \times 10^{-5}) \times 0.4}{5.01 \times 10^{-5}} = 1.08\,mol\,dm^{-3}$$

Step 5: Calculate number of moles: $1.08 \times \frac{250}{1000} = 0.27\,mol$

Determining the pH of a buffer solution

We can find the pH of a buffer solution experimentally by using a glass pH electrode connected to a pH meter. The glass pH electrode is dipped into the buffer solution and the value of the pH found by direct reading from the meter. The functioning of the glass electrode is based on the fact that the potential difference (voltage) between the surface of the glass and the solution varies linearly with the pH. A hydrogen electrode (see section 10.1) can also be used to determine pH.

Exam tips

In buffer solution calculations, do not fall into the trap of thinking that the concentration of hydrogen ions is the same as the concentration of conjugate base. Remember that hardly any of the conjugate base comes from ionisation of the acid. It is added as a salt.

Key points

- The pH of a buffer solution can be calculated using the equilibrium concentrations of the weak acid and its conjugate base together with the K_a value for the weak acid.
- The acid/ base ratio of a buffer solution can be calculated using the values of pH of the solution and the K_a value of the weak acid.
- The pH of a buffer solution can be determined experimentally using a glass electrode connected to a pH meter.

10 Electrode potentials and equilibrium

10.1 Introducing standard electrode potential

Learning outcomes

On completion of this section, you should be able to:

- define the term 'standard electrode potential'
- describe the standard hydrogen electrode
- describe standard electrode potentials for metal–metal ion systems.

Did you know?

The absolute potential between a metal and its ions in solution are thought to be caused by a build up of electronic charge on the surface of the metals. This attracts a layer of positive ions. An electrical double layer is formed. The more electrons given off by a metal, the greater, the charge difference between the metal and the metal ions in solution.

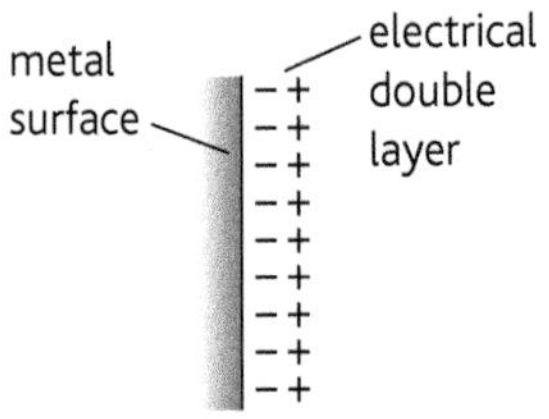

Introducing electrode potentials

When a copper rod is placed in a solution of Cu^{2+}(aq) ions, the following equilibrium is set up:

$$Cu^{2+}(aq) + 2e^- \rightleftharpoons Cu(s)$$

There are two opposing reactions:

- Cu^{2+} ions in solution accept electrons and become Cu atoms. The Cu^{2+} ions get reduced.
- Cu atoms lose electrons and become Cu^{2+} ions. The Cu atoms get oxidised.

For a relatively unreactive metal such as copper, the position of this equilibrium lies to the right. Cu^{2+} ions are relatively easy to reduce and Cu atoms are relatively difficult to oxidise.

For a relatively reactive metal such as magnesium, the position of this equilibrium lies to the left.

$$Mg^{2+}(aq) + 2e^- \rightleftharpoons Mg(s)$$

Mg^{2+} ions are relatively difficult to reduce and Mg atoms are relatively easy to oxidise.

When a metal is placed in a solution of its ions, a voltage (absolute potential) is established between the metals atoms and the metal ions in solution. It is not possible to measure this voltage directly. But we can measure the difference between one metal–metal ion system and another system. We call this value the **electrode potential**.

The standard hydrogen electrode

If we want to compare the ability of different metals to release electrons, we need a standard electrode for comparison. The electrode potential of the standard hydrogen electrode is by definition zero at all temperatures. The **standard hydrogen electrode** consists of:

- hydrogen gas at 101 kPa (1 atmosphere) in equilibrium with
- an aqueous solution of hydrogen ions at a concentration of 1.00 mol dm^{-3}
- a platinum electrode coated with platinum black in contact with both the hydrogen gas and hydrogen ions. (Platinum black is finely divided platinum which allows close contact between the hydrogen gas and the hydrogen ions.)

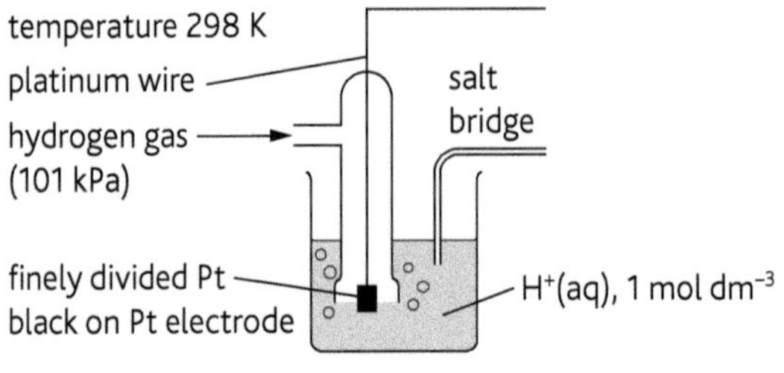

***Figure 10.1.1** The standard hydrogen electrode*

The half equation for this electrode is:

$$H^+(aq) + e^- \rightleftharpoons \frac{1}{2}H_2(g)$$

Measuring standard electrode potentials

In order to measure the electrode potential relating to the half equation:

$$Zn^{2+}(aq) + 2e^- \rightleftharpoons Zn(s)$$

we place a rod of pure zinc into a 1.00 mol dm^{-3} solution of Zn^{2+} ions. This is the Zn/Zn^{2+} **half cell**. We then connect this to a **standard hydrogen electrode**. We measure standard electrode potential, $E^{\ominus}$, by comparing the voltage of the half cell with the standard hydrogen electrode.

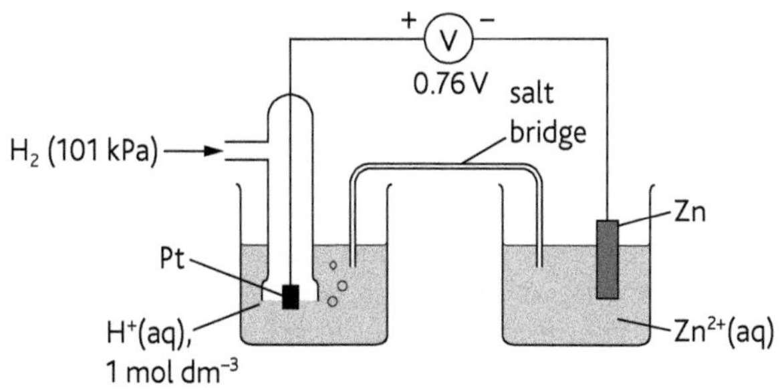

Figure 10.1.2 *Measuring* $E^{\ominus}$ *for a* Zn/Zn^{2+} *half cell*

The two half cells are connected by a **salt bridge**. This allows movement of ions between the two half cells, completing the electrical circuit so that the ionic balance is maintained. The salt bridge is a strip of filter paper or other inert porous material soaked in saturated potassium nitrate solution. The voltmeter should be a high resistance voltmeter.

Concentration of ions, pH of solution, temperature and pressure of gases all affect standard electrode potential. So we have to use standard conditions when comparing electrode potentials. These conditions are:

- temperature 25 °C (298 K)
- concentration of ions 1.00 mol dm^{-3}
- pressure of any gases involved 101 kPa.

The standard electrode potential, $E^{\ominus}$, of a half cell is the voltage measured under standard conditions with the standard hydrogen electrode as the other half cell.

For the Zn/Zn^{2+} half cell connected to the standard hydrogen electrode, the voltage developed is – 0.76 V. If we replace the Zn/Zn^{2+} half cell by a Cu/Cu^{2+} half cell, the voltage is + 0.34 V. The sign of the cell voltage depends on whether the half cell donates or receives electrons with respect to the standard hydrogen electrode (see Section 10.3).

Exam tip

- It is a common error to think that the salt bridge allows conduction of electrons. It connects two *ionic* solutions, so it must be allowing electrical conduction, maintaining *ionic* balance.
- Remember that the voltage of a cell is sometimes called its 'potential difference'.

Key points

- Electrode potential measures the ability of an element to release electrons or ions in solution to accept electrons.
- The standard hydrogen electrode is a half cell in which H_2 gas at a pressure of 101 kPa is in equilibrium with a solution of 1.00 mol dm^{-3} H^+ ions.
- The standard electrode potential, $E^{\ominus}$, of a half cell is the voltage the half cell under standard conditions compared with a standard hydrogen electrode.

10.2 Electrode potentials and cell potentials

Learning outcomes

On completion of this section, you should be able to:

- describe the measurement of standard electrode potentials for non-metal–non-metal ion systems
- describe the measurement of standard electrode potentials for systems involving ions of the same element in two different oxidation states
- define 'standard cell potential'
- calculate standard cell potential, $E^\ominus_{cell}$.

A variety of half cells

There are three main types of half cell whose $E^\ominus$ values can be obtained by connecting them to a standard hydrogen electrode:

- metal–metal ion half cells, the $Zn(s)/Zn^{2+}(aq)$ half cell (see Section 10.1)
- non-metal–non-metal ion half cells
- half cells containing ions of the same element in different oxidation states.

Half cells containing non-metals and non-metal ions

If no metal is present, electrical connection with the solution is made using a platinum electrode. The platinum is inert. It plays no part in the reaction but it must be in contact with both the element and an aqueous solution of its ions. Figure 10.2.1 shows the chlorine/ chloride ion half cell connected to the standard hydrogen electrode. The $E^\ominus$ value for this system is + 1.36 V.

$$\tfrac{1}{2}Cl_2(g) + e^- \rightleftharpoons Cl^-(aq)\ E^\ominus = +\ 1.36\,V$$

Note: It does not matter in terms of $E^\ominus$ values whether we write the half equation in this way or as $Cl_2(g) + 2e^- \rightleftharpoons 2Cl^-(aq)$

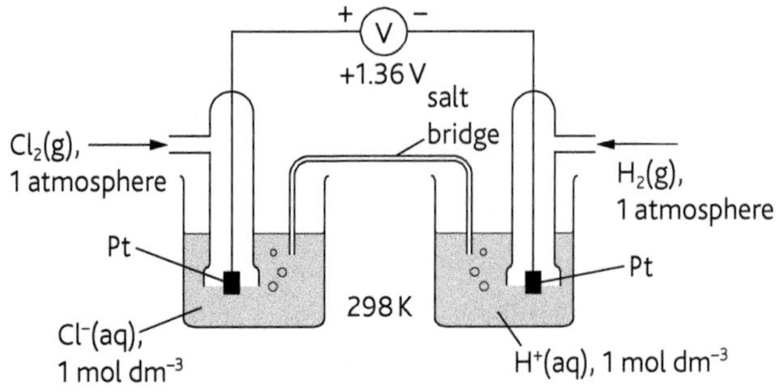

***Figure 10.2.1** Measuring $E^\ominus$ for the chlorine/ chloride half cell*

For a non-metallic liquid in equilibrium with its ions, e.g. bromine/ bromide, the platinum foil should be half in the bromine liquid and half in the bromide ions. The same applies to a solid in equilibrium with its ions, e.g. sulphur/ sulphide.

Half cells containing ions in different oxidation states

When a half cell contains ions of the same element in different oxidation states, both the ions must have the same standard concentration of $1.00\,mol\,dm^{-3}$. A platinum electrode is used to make electrical connection with the solution. Figure 10.2.2 shows the Fe^{3+}/Fe^{2+} half cell connected to the standard hydrogen electrode. The $E^\ominus$ value for this system is + 0.77 V.

$$Fe^{3+}(aq) + e^- \rightleftharpoons Fe^{2+}(aq)\ E^\ominus = +\ 0.77\,V$$

Some half cells may also contain acids or alkalis e.g.

$$MnO_4^-(aq) + 8H^+(aq) + 5e^- \rightleftharpoons Mn^{2+}(aq) + 4H_2O(l)\ E^\ominus = +\ 1.52\,V$$

The hydrogen ions are included because they are essential for the reaction. The MnO_4^-, Mn^{2+} and $8H^+$ are all present at a concentration of $1.00\,mol\,dm^{-3}$.

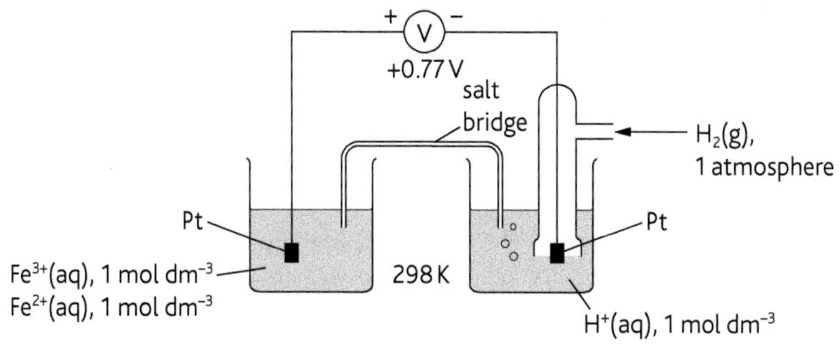

Figure 10.2.2 *Measuring* $E^\ominus$ *for the* Fe^{3+}/Fe^{2+} *half cell*

Standard cell potential

We can calculate the voltage of an electrochemical cell made up of two half cells whose $E^\ominus$ values relative to the standard hydrogen electrode are known.

The voltage measured is the difference between the $E^\ominus$ values of the two half cells. This is called the **standard cell potential**, $E^\ominus_{cell}$. The standard cell potential is the voltage developed under standard conditions when two standard half cells are joined.

Figure 10.2.3 shows a Cu/Cu^{2+} ion half cell connected to a Zn/Zn^{2+} half cell.

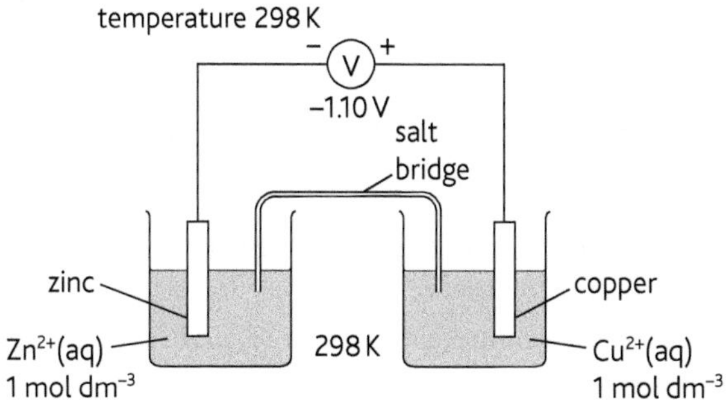

Figure 10.2.3 *Two half cells connected to give the standard cell potential*

The two relevant half equations are:

$$Cu^{2+}(aq) + 2e^- \rightleftharpoons Cu(s) \quad E^\ominus = +0.34\,V$$
$$Zn^{2+}(aq) + 2e^- \rightleftharpoons Zn(s) \quad E^\ominus = -0.76\,V$$

To find the value of $E^\ominus_{cell}$, we subtract the less positive (or more negative) $E^\ominus$ value from the more positive (or less negative) $E^\ominus$ value.

So in this case $E^\ominus_{cell} = +0.34 - (-0.76) = +1.10\,V$

Key points

- The standard electrode potential of a half cell is the voltage of the half cell compared with a standard hydrogen electrode.
- A platinum electrode is used with half cells in which there is no metal electrode.
- The standard cell potential is the voltage developed under standard conditions when two standard half cells are joined.
- Standard cell potential, $E^\ominus_{cell}$, is the difference between the $E^\ominus$ values of the two half cells.

Exam tips

Always make sure that you include the sign of electrode potential when you are doing calculations involving electrode potentials. The negatives and positive can get confusing.

10.3 Redox reactions and cell diagrams

Learning outcomes

On completion of this section, you should be able to:

- use standard electrode potentials to determine the direction of electron flow in a cell
- construct cell diagrams.

$E^\ominus$ values and redox reactions

- By convention, electrode potentials refer to a reduction reaction, so the electrons appear on the left of the half equations.
- The more positive (or less negative) the value of $E^\ominus$, the easier it is to reduce the ions or other species on the left of the equation. They are better at accepting electrons. So they are better oxidising agents.
- The more negative (or less positive) the value of $E^\ominus$, the easier it is to oxidise the species on the right of the half equation. They are better at releasing electrons. They are better reducing agents.

weaker oxidising agent — stronger reducing agent

increasing oxidising power of species in the left (gain e⁻ more readily) ↓	Half equation	$E^\ominus$	↑ increasing reducing power of species on the right (lose e⁻ more readily)
	$Cr^{2+}(aq) + 2e^- \rightleftharpoons Cr(s)$	$E^\ominus = -0.91\,V$	
	$Zn^{2+}(aq) + 2e^- \rightleftharpoons Zn(s)$	$E^\ominus = -0.76\,V$	
	$H^+(aq) + e^- \rightleftharpoons \frac{1}{2}H_2(g)$	$E^\ominus = 0.00\,V$	
	$Fe^{3+}(aq) + e^- \rightleftharpoons Fe^{2+}(aq)$	$E^\ominus = +0.77\,V$	
	$Ag^+(aq) + e^- \rightleftharpoons Ag(s)$	$E^\ominus = +0.80\,V$	
	$\frac{1}{2}Cl_2(g) + e^- \rightleftharpoons Cl^-(aq)$	$E^\ominus = +1.36\,V$	

stronger oxidising agent — weaker reducing agent

Referring to the list above, we see that:

- Cr and Zn are good reducing agents because they are more likely to release electrons (the reaction is further over to the left) compared with Cl^- ions or Ag.
- Cl_2 is a better oxidising agent than Fe^{3+} because it accepts electrons more readily than Fe^{3+} ions.
- Zn releases electrons more readily than H_2 and Fe^{3+} accepts electrons more readily than H^+.

The direction of electron flow in cells

When we connect two half cells:

- The half cell with the more positive (less negative) value of $E^\ominus$ is the positive pole.
- The half cell with the less positive (more negative) value of $E^\ominus$ is the negative pole.

So when the following two half cells are connected

Reaction 1: $Ag^+(aq) + e^- \rightleftharpoons Ag(s)$ $E^\ominus = +0.80\,V$
Reaction 2: $Cu^{2+}(aq) + 2e^- \rightleftharpoons Cu(s)$ $E^\ominus = +0.34\,V$

the Cu/Cu^{2+} half cell is the negative pole and the Ag/Ag^+ half cell is the positive pole.

This is because according to $E^\ominus$ values, copper is better at releasing electrons than silver and Ag^+ ions are better at accepting electrons than Cu^{2+} ions. Therefore Reaction 1 goes in the forward direct and Reaction 2 goes in the backward direction. So the direction of electron flow in the external wires is from the Cu/Cu^{2+} half cell to the Ag/Ag^+ half cell.

Exam tips

In order to deduce the direction of electron flow, remember that: the more *positive attracts the negative* electrons.

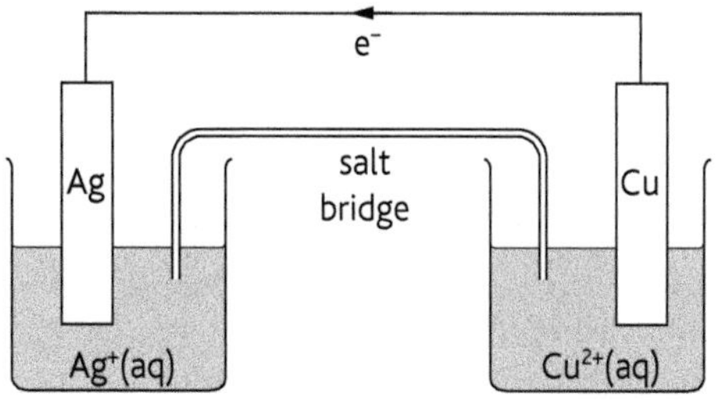

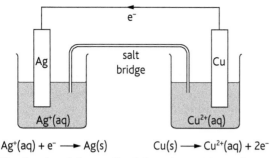

$Ag^+(aq) + e^- \longrightarrow Ag(s)$ $\quad\quad$ $Cu(s) \longrightarrow Cu^{2+}(aq) + 2e^-$

Figure 10.3.1 *The direction of electron flow is from the negative to the positive pole*

Cell diagrams

A convenient way of representing electrochemical cells is called a **cell diagram**. For a cell constructed from two metal/ metal ion systems we write the cell diagram as shown below. The reduced forms are on the outside of the diagram.

reduced form $\quad$ oxidised form $\quad$ oxidised form $\quad$ reduced form

$$Zn(s) \mid Zn^{2+}(aq) \parallel Cu^{2+}(aq) \mid Cu(s) \quad E^{\ominus} = +1.1\,V$$

phase change $\quad$ salt bridge

The sign of $E^{\ominus}$ in these diagrams represents the polarity (+ or –) of the right hand electrode in the diagram. In this case the Cu/Cu^{2+} half cell is + with respect to the Zn/Zn^{2+} half cell. If we write the cell diagram the other way round, the $E^{\ominus}$ is negative.

$$Cu(s) \mid Cu^{2+}(aq) \parallel Zn^{2+}(aq) \mid Zn(s) \quad E^{\ominus} = -1.1\,V$$

The standard half cell diagram for the hydrogen electrode is:

$$Pt\ [H_2(g)] \mid 2H^+(aq) \parallel$$

The hydrogen gas and platinum electrode are regarded as being a single phase.

For more complex half cells which involve non-metals or ions in different oxidation states, the platinum is shown as a separate phase from the aqueous ions and a comma separates the different oxidation states e.g.

$$Pt \mid Fe^{2+}(aq), Fe^{3+}(aq) \parallel Cu^{2+}(aq) \mid Cu(s)\ E^{\ominus} = -0.43\,V$$

If hydrogen or hydroxide ions appear in the half equation, they are bracketed in the cell diagram, e.g.

$$Pt \mid [Mn^{2+}(aq) + 4H_2O(l)], [MnO_4^-(aq) + 8H^+(aq)] \parallel Ag^+(aq) \mid Ag(s)$$

reduced form $\quad\quad$ oxidised form

Key points

- The direction of electron flow in a cell is from the half cell with the most negative (less positive) electrode potential to the half cell with the less negative electrode potential.
- Cell diagrams show each of the oxidised and reduced species and other ions which may take part in the reaction.
- In a cell diagram the reduced species are placed on the outside and the oxidised species on the inside.

10.4 Using $E^\ominus$ values to predict chemical change

Learning outcomes

On completion of this section, you should be able to:

- use $E^\ominus$ values to predict whether a reaction is likely to occur or not.

Using $E^\ominus$ values to predict a reaction

Figure 10.4.1 compares some oxidising and reducing powers of some elements and ions by comparing $E^\ominus$ values:

species on left gain electrons more readily (↓)

$Mg^{2+}(aq) + 2e^- \rightleftharpoons Mg(s)$	$E^\ominus = -2.38\,V$
$Zn^{2+}(aq) + 2e^- \rightleftharpoons Zn(s)$	$E^\ominus = -0.76\,V$
$H^+(aq) + e^- \rightleftharpoons \frac{1}{2}H_2(g)$	$E^\ominus = 0.00\,V$
$Cu^{2+}(aq) + 2e^- \rightleftharpoons Cu(s)$	$E^\ominus = +0.34\,V$
$Fe^{3+}(aq) + e^- \rightleftharpoons Fe^{2+}(aq)$	$E^\ominus = +0.77\,V$
$Ag^+(aq) + e^- \rightleftharpoons Ag(s)$	$E^\ominus = +0.80\,V$
$\frac{1}{2}Cl_2(g) + e^- \rightleftharpoons Cl^-(aq)$	$E^\ominus = +1.36\,V$

species on right lose electrons more readily (↑)

***Figure 10.4.1** As the value of $E^\ominus$ gets more positive the species on the left are stronger oxidising agents and the species on the right are weaker reducing agents*

- For each half equation, the more oxidised form is on the left.
- The more positive the value of $E^\ominus$, the greater is the tendency of the reaction to move in the forward direction. This is because it is better at accepting electrons than the species above it. It is a better oxidising agent.
- The more negative the value of $E^\ominus$, the greater is the tendency of the reaction to move in the reverse direction. This is because it is better at losing electrons than the species below it. It is a better reducing agent.

We can use these ideas to predict whether a reaction will take place or not. If a reaction is likely to take place using information from $E^\ominus$ values, we say that it is **feasible**. For example: Will zinc react with copper(II) ions?

The relevant half reactions are:

$$Zn^{2+}(aq) + 2e^- \rightleftharpoons Zn(s) \quad E^\ominus = -\,0.76\,V$$
$$Cu^{2+}(aq) + 2e^- \rightleftharpoons Cu(s) \quad E^\ominus = +\,0.34\,V$$

Cu^{2+} ions have a greater tendency to gain electrons than Zn^{2+} ions and Zn has a greater tendency to lose electrons than Cu. So the position of equilibrium of the Zn/ Zn^{2+} reaction moves to the left and the position of equilibrium of the Cu/Cu^{2+} reaction moves to the right. The overall reaction is:

$$Zn(s) + Cu^{2+}(aq) \rightarrow Zn^{2+}(aq) + Cu(s)$$

In these predictions, we are assuming that the standard conditions such as occur in a standard electrochemical cell apply. If standard conditions do not apply, we have to consider other information (see Section 10.5).

The anticlockwise rule

We can predict whether a reaction is feasible by:

- writing down the two half reactions with the one with the most negative value at the top
- the direction in which the reaction is feasible is given an anticlockwise pattern starting from the bottom left.
- the pattern from the bottom left is: reactant → product → reactant → product

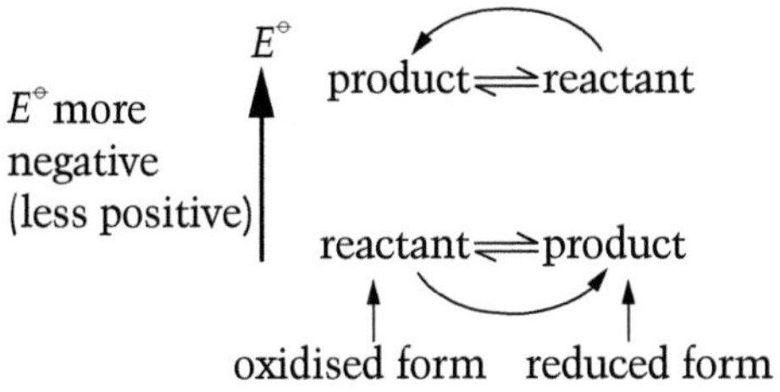

Example 1

Will chlorine oxidise Fe^{2+} ions to Fe^{3+} ions?

$E^\ominus$				
+0.77	$Fe^{3+}(aq)$	+ e−	⇌	$Fe^{2+}(aq)$ better reducing agent
+1.36	$\frac{1}{2}Cl_2(g)$ better oxidising agent	+ e−	⇌	$Cl^-(aq)$

You can see that the better reducing agent reacts with the better oxidising agent. Applying the anticlockwise rule, the reaction is feasible:

$$\tfrac{1}{2}Cl_2(g) + Fe^{2+}(aq) \rightarrow Cl^-(aq) + Fe^{3+}(aq)$$

This because chlorine is better at accepting electrons than Fe^{2+} ions and Fe^{3+} is better at releasing electrons than Cl^- ions.

Example 2

Will aqueous iodine oxidise silver to silver ions Ag^+?

$E^\ominus$				
+0.54	$\frac{1}{2}I_2(aq)$	+ e−	⇌	$I^-(aq)$ better reducing agent
+0.80	$Ag^+(aq)$ better oxidising agent	+ e−	⇌	$Ag(aq)$

Applying the anticlockwise rule, the reaction is not feasible. The reaction will be between silver ions and iodide ions rather than between silver and iodide ions.

Warning! The feasibility of a reaction based on $E^\ominus$ values only tells us that a reaction is possible. It is no guarantee that a reaction will happen. Some reactions, although feasible, take place so slowly that they do not seem to be happening. We say that these reactions are kinetically controlled.

Selecting $E^\ominus$ values

You must take care when selecting $E^\ominus$ values from a table of data. Some half reactions may look very similar, especially if two ions are involved e.g.

$$VO^{2+} + 2H^+ + e^- \rightleftharpoons V^{3+} + H_2O \quad E^\ominus = +0.34\,V$$
$$VO_2^+ + 2H^+ + e^- \rightleftharpoons VO^{2+} + H_2O \quad E^\ominus = +1.00\,V$$

Exam tips

If you arrange the two half reactions with the one with the most negative value at the top, the reactants will be the bottom left and top right species.

Key points

- A reaction is feasible if the $E^\ominus$ value for the reaction in the forward direction is positive.
- A reaction is feasible if the $E^\ominus$ value of the half equation involving the species being reduced is more positive than the $E^\ominus$ value of the half equation of the species being oxidised.
- The direction of a reaction can be predicted from diagrams showing half equations and their $E^\ominus$ values.

10.5 Electrode potentials and electrochemical cells

Learning outcomes

On completion of this section, you should be able to:

- describe how $E^\ominus$ values vary with concentration
- know how to apply the principles of redox equilibria to energy storage devices including fuel cells.

Concentration change and electrode potential

The position of equilibrium is affected by changes in concentration and temperature. This also applies to redox equilibria. $E^\ominus$ values refer to a concentration of ions of 1.00 mol dm^{-3} and temperature of 298 K. What happens when we alter the concentration of ions in a metal–metal-ion system? Take for example the half reaction:

$$Cr^{2+}(aq) + 2e^- \rightleftharpoons Cr(s) \quad E^\ominus = -0.91\,V$$

- If the concentration of $Cr^{2+}(aq)$ increases, the value of E becomes less negative. e.g. –0.83 V.
- If the concentration of $Cr^{2+}(aq)$ decreases, the value of E becomes more negative. e.g. –0.99 V.

We can apply Le Chatelier's principle to concentration changes in redox reactions such as this one:

- If the concentration of the ion on the left of the equation increases, the equilibrium shifts to the right as more electrons are accepted. If electrons are more easily accepted, the value of E gets more positive (or less negative).
- If the concentration of the ion on the left of the equation decreases, the equilibrium shifts to the left as more electrons are released. If electrons are more easily released, the value of E gets more negative (or less positive).

For a system such as $Fe^{3+}(aq) + e^- \rightleftharpoons Fe^{2+}(aq)$, increasing (or decreasing) the concentrations of both ions equally has no effect on the value of E. This is because the increase in the concentration of Fe^{3+} pushes the position of equilibrium to the right, but increase in the concentration of Fe^{2+} pushes the position of equilibrium to the left to an equal extent.

Predicting a reaction under non-standard conditions

If reactions are carried out under non-standard conditions it is more difficult to predict if a reaction is feasible. As a rough guide we can say that:

- If the electrode potentials of two half reactions differ by more than 0.3 V, the predicted reaction is likely to happen, e.g. the reaction between copper and Fe^{3+} ions is likely to be feasible because the difference is 0.43 V.

$$Cu^{2+}(aq) + 2e^- \rightleftharpoons Cu(s) \quad E^\ominus = +0.34\,V$$
$$Fe^{3+}(aq) + e^- \rightleftharpoons Fe^{2+}(aq) \quad E^\ominus = +0.77\,V$$

- If the difference in electrode potentials of two half reactions differ by less than 0.3 V, the feasibility of the reaction may not be predicted with confidence, e.g. the reaction between Ag^+ ions and Fe^{2+} ions cannot be predicted because the difference is 0.03 V.

$$Fe^{3+}(aq) + e^- \rightleftharpoons Fe^{2+}(aq) \quad E^\ominus = +0.77\,V$$
$$Ag^+(aq) + e^- \rightleftharpoons Ag(s) \quad E^\ominus = +0.80\,V$$

Batteries and cells

Most batteries and cells we use do not operate under standard conditions. They often have higher concentrations of electrolyte. Modern batteries and cells have many advantages:

- They are often small in size and lightweight. Car batteries, however are quite heavy, although lightweight ones are being developed.
- They deliver a high voltage for a considerable period of time.
- Some can be recharged.

Button cells

Many button cells such as those used to power a watch deliver a high voltage. One type of cell relies on the reaction of lithium and iodine. It is a solid state cell so there is no spillage of liquid electrolyte. Most button cells do not contain a liquid electrolyte or the electrolyte is a paste.

The half equations are:

$$Li^+ + e^- \rightleftharpoons Li \quad E^\ominus = -3.14\,V$$
$$\tfrac{1}{2}I_2 + e^- \rightleftharpoons I^- \quad E^\ominus = +0.54\,V$$

The voltage of this cell should be $+0.54 - (-3.14) = 3.58\,V$.

The voltage of the cell in use, however, is more likely to be about 3 V because the conditions are not standard.

Another type of button cell is the zinc–silver oxide cell. The zinc is the negative pole and the silver oxide is the positive pole.

$$Zn(s) + Ag_2O(s) \rightarrow ZnO(s) + 2Ag \quad E^\ominus_{cell} = +1.60\,V$$

Fuel cells

In a fuel cell, a fuel (often hydrogen) releases electrons at one electrode and oxygen gains electrons at the other electrode. The relevant $E^\ominus$ values for a fuel cell with an acidic electrolyte are:

$$2H^+ + 2e^- \rightleftharpoons H_2 \quad E^\ominus = 0.00\,V$$
$$4H^+ + O_2 + 4e^- \rightleftharpoons 2H_2O \quad E^\ominus = +1.23\,V$$

The overall reaction is: $2H_2 + O_2 \rightarrow 2H_2O$.

The fuel cells used in some cars and buses have several advantages over petrol or diesel engines:

- Water is the only product. No carbon dioxide or nitrogen oxides are released.
- They produce more energy per gram of fuel than petrol or diesel.
- They are very efficient. The transmission of power to the engine is more direct.

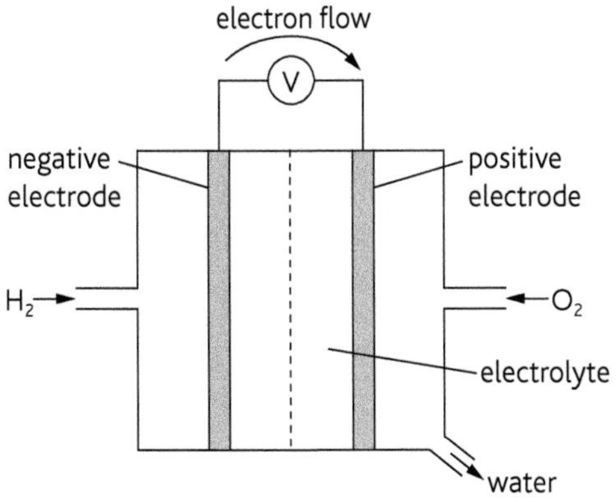

Figure 10.5.1 *Simplified diagram of a hydrogen–oxygen fuel cell*

Did you know?

The use of fuel cells have many disadvantages as well as advantages:

- The storage of hydrogen in large quantities is a problem.
- Refuelling has to be carried out more often than with a petrol engine.
- The hydrogen is produced from fossil fuels.
- They do not work well at low temperatures.
- They are expensive.

Key points

- The value of $E^\ominus$ in a metal/metal ion half cell becomes more negative as the concentration of the ions decreases.
- Modern batteries have advantages in terms of small size, low mass and high voltage.
- Fuel cells generate a voltage by using the energy from the reaction of oxygen with a fuel.

Revision questions

1 For each reaction given, write the specified equilibrium expression and state the associated units. Where K_p is required, assume the partial pressures are measured in atmospheres.

a i $2NO_2Cl(g) \rightleftharpoons 2NO_2(g) + Cl_2(g)$; K_c

ii $Fe^{3+}(aq) + I^-(aq) \rightleftharpoons Fe^{2+}(aq) + \frac{1}{2}I_2(aq)$; K_c

b i $3O_2(g) \rightleftharpoons 2O_3(g)$; K_p

ii $NH_3(g) + ClF_3(g) \rightleftharpoons 3HF(g) + \frac{1}{2}N_2(g) + \frac{1}{2}Cl_2(g)$; K_p

c i $Ba(IO_3)_2(s) \rightleftharpoons Ba^{2+}(aq) + 2IO_3^-(aq)$; K_{sp}

ii $Al(OH)_3(s) \rightleftharpoons Al_3^+(aq) + 3OH^-(aq)$; K_{sp}

d i $HF(aq) + H_2O(l) \rightleftharpoons H_3O^+(aq) + F^-(aq)$; K_a

ii $HCN(aq) \rightleftharpoons H^+(aq) + CN^-(aq)$; K_a

e $HS^-(aq) + H_2O(l) \rightleftharpoons H_2S(aq) + OH^-(aq)$; K_b

2 a What do you understand by the term 'dynamic equilibrium'?

b Based on Le Chatelier's principle, explain how each of the factors listed would affect the equilibrium system.

$PCl_5(g) \rightleftharpoons PCl_3(g) + Cl_2(g)$; $\Delta H = +87.9$ kJ

(Consider yield, position of equilibrium and value of equilibrium constant.)

i adding more PCl_3

ii decreasing the pressure

iii adding a catalyst

iv increasing the temperature

v removing Cl_2?

c For the reaction $NH_3(aq) + H^+(aq) \rightleftharpoons NH_4^+(aq)$, where the forward reaction is exothermic, what would be the effect of:

i increasing the pressure

ii increasing the concentration of $NH_3(aq)$

iii increasing the temperature?

3 a For the system $2NO_2(g) \rightleftharpoons 2NO(g) + O_2(g)$, which has reached equilibrium in a 2 dm^3 flask containing 0.0222 mol NO_2, 6.34 g NO and 10.8 g O_2, what would be the value of K_c, along with the appropriate units?

b For the system $A(aq) + B(aq) \rightleftharpoons R(aq)$, at a particular temperature, $K_c = 4.5 \times 10^{-2}\,dm^3\,mol^{-1}$. In an equilibrium mixture at the same temperature, the known concentrations in mol dm^{-3} are [A] = 2.10, [B] = 1.45. Calculate the equilibrium concentration of R.

c A mixture of 1.40 mol H_2 and 1.40 mol I_2 are placed in a 2 dm^3 flask at constant temperature, and the equilibrium $H_2(g) + I_2(g) \rightleftharpoons 2HI(g)$ is established. The equilibrium mixture contains 0.36 mol H_2. What is the value of K_c at this temperature?

d When 6.00 mol SO_2Cl_2 are placed in a 3 dm^3 flask at a high temperature, 40% decomposes to SO_2 and Cl_2 as shown by the equation:

$$SO_2Cl_2(g) \rightleftharpoons SO_2(g) + Cl_2(g)$$

Calculate K_c at this temperature.

4 a A mixture of 6.5 g H_2 and 4.0 g N_2 are placed in a 5.0 dm^3 vessel. If the total pressure is 3.9 atm, what are the partial pressures of N_2 and H_2?

b For the equilibrium system $2NO(g) + Cl_2(g) \rightleftharpoons 2NOCl(g)$, an equilibrium mixture at a constant high temperature has partial pressures of NO = 9.6×10^3 Pa, $Cl_2 = 1.7 \times 10^4$ Pa, NOCl = 2.8×10^4 Pa. Calculate K_p at this temperature.

c At a particular temperature, a vessel initially contains only N_2 at a partial pressure of 2.50 atm and O_2 at a partial pressure of 1.50 atm. The equilibrium $N_2(g) + O_2(g) \rightleftharpoons 2NO(g)$ is established and it is found that the mixture contains NO at a partial pressure of 0.76 atm. What is the value of K_p?

5 a What is meant by the term solubility?

b The solubilities in pure water, of the substances given in **i** and **ii** below were determined at 25 °C. What is the K_{sp} at this temperature for each substance?

i 0.710 g of $AgBrO_3$ in 400 ml of water.

ii 7.52×10^{-3} mol dm^{-3} BaF_2.

c i The solubility product, K_{sp} for $MgCO_3(s)$ is $3.5 \times 10^{-8}\,mol^2\,dm^{-6}$ at 25 °C. What would be the solubility, at this temperature in

- pure water,
- a solution containing 0.015 mol dm^{-3} Mg^{2+} ions?

ii Explain why the solubilities in **i** differ.

iii Would $MgCO_3$ precipitate if a solution containing 0.055 mol dm^{-3} magnesium nitrate is mixed with a solution containing 1.22×10^{-3} mol dm^{-3} sodium carbonate? Explain your answer.

6 **a** The concentrations of HCl and C_6H_5COOH are both 0.025 mol dm^{-3}.

- **i** Why is the pH different for these two acids, even though they have the same concentration?
- **ii** What is the pH of the HCl?
- **iii** What is the pH of the benzoic acid at 25 °C, given the K_a at this temperature is 6.3×10^{-5} mol dm^{-3}?

b What is the pH of a 0.021 mol dm^{-3} solution of $HClO_4$? What is the pOH?

c The pH of a solution was found to be 11.4. What is

- **i** $[H^+]$,
- **ii** $[OH^-]$?

7 **a** **i** Write a suitable equation to show how the base phenylamine, ($C_6H_5NH_2$), dissociates in water.

- **ii** What is the conjugate acid of this base?
- **iii** The K_b for $C_6H_5NH_2$ is 4.3×10^{-10} mol dm^{-3}. Calculate
  - the pH of a 0.225 mol dm^{-3} solution,
  - the pK_b.

b **i** Explain why a mixture of $HC_3H_5O_3$ (lactic acid) and $NaC_3H_5O_3$ (sodium lactate) would function as a buffer whereas a mixture of nitric acid and sodium nitrate would not.

- **ii** The K_a for CH_3COOH is 1.8×10^{-5} mol dm^{-3}. What is the pH of a buffer solution formed by mixing 15.5 g of CH_3COOH (ethanoic acid) with 15.5 g of CH_3COONa (sodium ethanoate) in a 250 cm^3 volumetric flask?
- **iii** The K_b for CO_3^{2-} is 2.1×10^{-4} mol dm^{-3}. What is the pH of a buffer formed by mixing 0.105 mol dm^{-3} Na_2CO_3 with 0.120 mol dm^{-3} $NaHCO_3$?

8 A volume of 30 cm^3 of 0.15 mol dm^{-3} KOH solution was placed in a conical flask and a total volume of 70 cm^3 of 0.10 mol dm^{-3} HNO_3 solution added in 5.0 ml aliquots. The equivalence volume was found to be 45 cm^3.

- **a** What is the initial pH of the KOH solution?
- **b** What is the pH of the solution after the addition of 20 ml of HNO_3?
- **c** What is the pH of the solution after adding 55 cm^3 of HNO_3?

9 **a** Draw a diagram of the apparatus which would be used to determine the standard electrode potential of the half cell Sn^{4+}(aq)/Sn^{2+}(aq).

b **i** Draw a cell diagram for the combination of the two half cells Al^{3+}(aq)/Al(s); $E^\ominus = -1.66$ V and Zn^{2+}(aq)/Zn(s); $E^\ominus = -0.76$ V

- **ii** Calculate the standard cell emf for this cell.
- **iii** Write the balanced equation for the cell reaction.

Exam-style questions – Module 2

Answers to all exam-style questions can be found on the accompanying CD

Multiple-choice questions

1 A radioactive source has 1.6×10^{20} atoms of a radioactive isotope, with a half-life of 3 days. How many atoms will decay in 12 days?

A 1.0×10^{19} **C** 4.0×10^{19}

B 1.2×10^{20} **D** 1.5×10^{20}

2 For the reaction X + Z → M, the concentration of reactants and products is measured in mol dm^{-3} and time is measured in seconds. The rate equation is

$$\text{Rate} = k[Z]^2$$

What conclusion can be made from this information?

A The unit for the rate constant, k, is dm^3 mol^{-1} s^{-1}.

B The unit for the rate constant is mol dm^{-3} s^{-1}.

C The reaction rate doubles, if the concentration of Z doubles.

D The reaction rate increases, as the concentration of reactant X increases.

3 For the reaction, $NO_2(g) + CO(g) \rightarrow NO(g) + CO_2(g)$, the following two-step mechanism is proposed:

Step 1: $NO_2(g) + NO_2(g) \rightarrow NO_3(g) + NO(g)$ *slow*

Step 2: $NO_3(g) + CO(g) \rightarrow NO_2(g) + CO_2(g)$ *fast*

What is the rate equation for this reaction?

A Rate = $k[NO_3][CO]$

B Rate = $k[NO_2]^2$

C Rate = $k[NO_3][NO]$

D Rate = $k[NO_2][CO_2]$

4 Three monobasic acids X, Y and Z all have a concentration of 0.011 mol dm^{-3}. The pK_a of X is 5.61, the K_a of Y is 4.13×10^{-7} and the pH of Z is 4.78. What is the order of the acids in terms of increasing acid strength?

A X < Y < Z

B Z < X < Y

C Y < X < Z

D Z < Y < X

5 Given the following half cells and corresponding standard electrode potential values,

half cell	$E^\ominus$/V
$Mn^{3+}(aq) + e^- \rightleftharpoons Mn^{2+}(aq)$	+1.49
$Zn^{2+}(aq) + 2e^- \rightleftharpoons Zn(s)$	−0.76

what is the correct cell diagram and $E^\ominus_{cell}$?

A $Zn(s)\,|\,Zn^{2+}(aq)\,||\,Mn^{3+}(aq)\,|\,Mn^{2+}(aq)$ +0.73 V

B $Mn^{2+}(aq)\,,\,Mn^{3+}(aq)\,||\,Zn^{2+}(aq)\,|\,Zn(s)$ −1.25 V

C $Pt(s)\,|\,Mn^{2+}(aq), Mn^{3+}(aq)\,||\,Zn\,(s)\,|\,Zn^{2+}(aq)$ +0.73 V

D $Zn(s)\,|\,Zn^{2+}(aq)\,||\,Mn^{3+}(aq), Mn^{2+}(aq)\,|\,Pt(s)$ +2.25 V

6 From the data given, which of the following are true?

	$E^\ominus$
$Sn^{4+}(aq)/Sn^{2+}(aq)$	+0.15 V
$V^{2+}(aq)/V(s)$	−1.2 V
$MnO_4^-(aq)/MnO_4^{2-}(aq)$	+0.56 V

i MnO_4^- is a stronger oxidising agent than $Sn^{4+}(aq)$.

ii V will reduce Sn^{4+} to Sn^{2+}.

iii A reaction between Sn^{2+} and V^{2+} is feasible.

A **i**, **ii** and **iii** are true.

B Only **i** and **ii** are true.

C Only **i** and **iii** are true.

D Only **ii** is true.

7 Consider the reaction,

$$X_2(g) + 2Y_2(g) \rightleftharpoons 2XY_2(g);\ \Delta H = +58.0\text{ kJ}$$

Which of the following combinations correctly predicts the effect of increasing the temperature?

	Position of equilibrium	*Yield*	*Value of K_c*
A	shift to left	decreases	no change
B	shift to right	increases	increases
C	shift to right	decreases	decreases
D	shift to left	increases	decreases

8 For propanoic acid, which has $K_a = 1.35 \times 10^{-5}$ mol dm^{-3} the dissociation in water is

$$CH_3CH_2COOH(aq) + H_2O(l) \rightleftharpoons CH_3CH_2COO^-(aq) + H_3O^+(aq)$$

Which of the following statements are true?

i The $K_a = \dfrac{[CH_3CH_2COO^-(aq)][H_3O^+(aq)]}{[CH_3CH_2COOH(aq)][H_2O(l)]}$

ii The pH of a 0.010 mol dm^{-3} solution is 2.

iii Adding OH^- would cause the equilibrium to shift to the right.

iv Increasing the molarity of the acid would result in a higher concentration of H_3O^+ ions at equilibrium.

A **i**, **ii** and **iii** and **iv** are true.

B **ii**, **iii** and **iv** are true.

C **iii** and **iv** are true.

D **i** and **ii** are true.

9 What is the solubility of $Zn(OH)_2$ in pure water at 25 °C, if the solubility product constant, K_{sp} under these conditions is 3.0×10^{-16} mol^3 dm^{-9}?

A 6.7×10^{-6} mol dm^{-3}

B 1.7×10^{-8} mol dm^{-3}

C 5.3×10^{-6} mol dm^{-3}

D 4.2×10^{-6} mol dm^{-3}

10 The solubility of copper(I) bromide CuBr in pure water at 25 °C is 7.28×10^{-5} mol dm^{-3}. Which of the following conclusions are correct, based on this data?

i The solubility will be greater than 7.28×10^{-5}, if CuBr is dissolved in a solution of KBr.

ii The numerical value of the K_{sp} is 5.3×10^{-9}.

iii The unit of the K_{sp} would be mol^2 dm^{-6}.

iv The concentration of the Cu^+ ion at equilibrium, in pure water, would be 7.28×10^{-5} mol dm^{-3}.

A **i**, **ii** and **iii** and **iv** are true.

B **ii**, **iii** and **iv** are true.

C **ii** and **iii** are true.

D **i**, **ii** and **iii** are true.

Structured questions

11 A student pipetted 25.0 ml of 0.112 mol dm^{-3} of a weak monobasic acid solution into a conical flask. To this was added an alkaline solution, X, from a burette. After each addition, the solution was swirled to ensure complete mixing and the pH was then measured using a pH meter. The student's results are shown on the graph below:

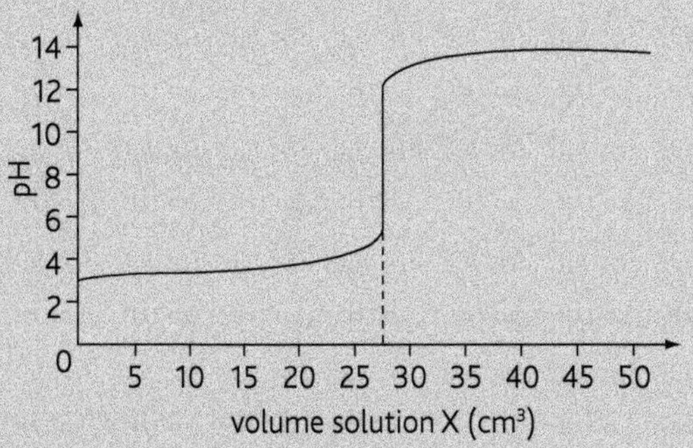

a Give the name or formula of an alkali that could produce the results shown on the graph. [1]

b **i** What is the K_a of the acid? [3]

ii Determine the pK_a for the acid. [1]

c From the equivalence volume and any other relevant data, determine the molarity of the base. [3]

d Explain why the pH at the equivalence point is greater than 7. [2]

e The solution has its maximum buffering capacity at $\frac{V}{2}$ where V is the equivalence volume.

i What is the pH at this point? [1]

ii What do you understand by the term 'buffering capacity'? [2]

f The indicator phenol red has a pH range of 6.8–8.4. State, giving your reasoning, whether it is suitable for use in a this titration. [2]

12 This question is based on the two half cells given below, along with their standard electrode potential values.

$$Ag^+(aq) + e^- \rightleftharpoons Ag(s) \qquad E^\ominus = +0.80\,V$$

$$MnO_4^-(aq) + 8H^+(aq) + 5e^- \rightleftharpoons Mn^{2+}(aq) + 4H_2O(l) \qquad E^\ominus = +1.52\,V$$

a What do you understand by the term 'standard electrode potential'? [2]

b **i** Draw a labelled diagram of the apparatus, which could be used to set up this cell, in order to determine the standard cell e.m.f. [3]

ii On the diagram,

- Label the anode and the cathode. [1]
- Show the direction of electron flow. [1]

c Calculate the standard cell e.m.f. and write a balanced equation for the cell reaction. [2]

d Write the cell diagram. [2]

e How would changing the concentration of the solutions in the respective half cells affect the value of the cell electromotive force? [2]

f Explain why KCl would not be suitable for use in the salt bridge. [2]

13 **a** **i** State the postulates of the collision theory. [2]

ii Based on the collision theory, explain why increasing the concentration of a reactant would generally cause an increase in the rate of a reaction. [2]

b The results shown in the table below were obtained for the reaction

$$A(aq) + B(aq) \rightarrow C(aq)$$

where the initial rate was determined by varying concentrations of A and B, at room temperature.

Experiment	[A] (mol dm^{-3})	[B] (mol dm^{-3})	Initial rate (mol dm^{-3} s^{-1})
1	0.125	0.125	1.45×10^{-2}
2	0.250	0.125	2.90×10^{-2}
3	0.125	0.250	5.80×10^{-2}

i Sketch a graph to illustrate how the initial rate of a reaction could be determined from monitoring the change in concentration of a reactant over time. [2]

ii Giving your reasoning, determine

- the order with respect to A [2]
- the order with respect to B. [2]

iii Write the rate equation for the reaction. [1]

iv Calculate a value for the rate constant, k, at this temperature, and state the unit. [2]

v What would be the rate of reaction, if the starting concentration of A was 0.30 mol dm^{-3} and B was 0.14 mol dm^{-3}? [2]

11 The elements of Period 3

11.1 Physical properties of the Period 3 elements

Learning outcomes

On completion of this section, you should be able to:

- explain the variation in some physical properties of Period 3 elements in terms of structure and bonding with reference to melting points, electrical conductivity, electronegativity and density.

Did you know?

Sulphur exists in different forms (**allotropes**). These different forms have different shaped crystals and slightly different melting points. The melting point of monoclinic sulphur (which has rod-like crystals) is 392 K, but the melting point of rhombic sulphur (which has octahedral shaped crystals) is 386 K. Phosphorus also has red, white and black forms. It is the white form which melts at 317 K.

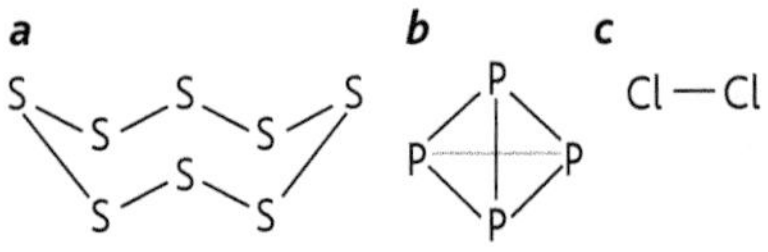

Figure 11.1.2 *The structures of:* ***a*** *sulphur;* ***b*** *phosphorus and;* ***c*** *chlorine*

Structure and melting points

The Period 3 elements, their physical state at room temperature and melting points in kelvin are shown below:

Group							
I	II	III	IV	V	VI	VII	0
sodium	magnesium	aluminium	silicon	phosphorus	sulphur	chlorine	argon
371 K	922 K	933 K	1683 K	317 K	392 K	172 K	84 K
solid	solid	solid	solid	solid	solid	gas	gas

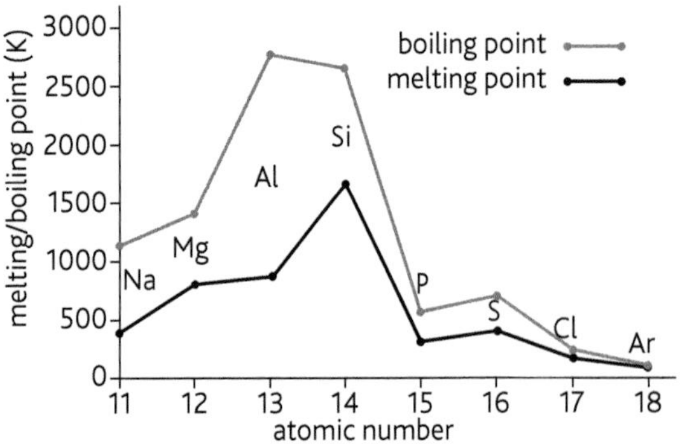

Figure 11.1.1 *Melting and boiling points of the Period 3 elements*

We can explain the differences in melting point in terms of structure and bonding:

- Sodium, magnesium and aluminium have a giant metallic structure. The ions are held together by a sea of delocalised electrons. From sodium to aluminium there is an increase in the number of electrons donated to the sea of electrons when the metal ions Na^+, Mg^{2+} and Al^{3+} form. The greater the ionic charge and number of delocalised electrons, the greater is the electrostatic attraction between the ions and the electrons and the more difficult it is to overcome these forces. So the melting points and boiling points increase in the order Na to Mg to Al.
- Silicon, in Group IV has the highest melting point of the Period 3 elements. It forms a giant covalent lattice similar to diamond. It takes a lot of energy to break the large number of strong covalent bonds. So the melting point is very high.
- Phosphorus, sulphur and chlorine have simple molecular structures. They have low melting points because there are only weak van der Waals attractive forces between the molecules. The van der Waals forces increase with increasing number of electrons in the molecule and with the number of contact points between neighbouring molecules. So sulphur, S_8, has a higher melting point than phosphorus, P_4. Chlorine is a gas since it is a diatomic molecule with a fairly small number of electrons.
- Argon exists only as isolated atoms. The van der Waals forces between these are very low, so it has a very low melting point.

Density of Period 3 elements

The table below shows the density of the Period 3 elements. The density of Cl and Ar are those of their liquids. They are gases at room temperature so have very low densities.

Element	Na	Mg	Al	Si	P	S	Cl	Ar
Density/$g\,cm^{-3}$	0.97	1.74	2.70	2.32	1.82	2.07	1.56	1.40

Density depends on: (i) the mass of the atoms, (ii) the size of the atoms, (iii) the way the atoms are packed in their lattice.

- The density increases from Na to Al as the size of the atoms decreases and their mass increases.
- Si has a lower density than Al because it has a more open lattice structure.
- The density of P and S (and Cl and Ar as liquids) is relatively low. Although the atoms are heavier, the way the molecules are packed together has an effect.

Electrical conductivity of Period 3 elements

The electrical conductivity (measured in siemens per metre, $S\,m^{-1}$) of the Period 3 elements are shown below. Chlorine and argon are gases and hardly conduct as solids.

Element	Na	Mg	Al	Si	P	S
Conductivity/ $10^8\,S\,m^{-1}$	0.218	0.224	0.382	2×10^{-10}	1×10^{-17}	1×10^{-23}

Na, Mg and Al are all good conductors. The delocalised electrons in the metallic structure are free to move when a voltage is applied. The conductivity increases from Na to Al as the number of delocalised electrons provided by each 'atom' increases (Na provides 1 electron, Mg provides 2 and Al provides 3). The electrical conductivity of Si is poor because it has no delocalised electrons. It is classed as a **metalloid** (a substance whose electrical conductivity increases with increase in temperature). Some of its electrons, however, can move out of position, especially when 'contaminating' atoms are present in its lattice. So it is called a **semi-conductor**. P and S hardly conduct electricity because they are covalent molecules with no delocalised electrons.

Electronegativity of Period 3 elements

The electronegativity of the atoms increases across Period 3.

Element	Na	Mg	Al	Si	P	S	Cl
Electronegativity	0.9	1.2	1.5	1.8	2.1	2.5	3.0

Key points

- Across Period 3 the structure of the elements changes from metallic to giant covalent to simple molecular to atomic.
- The variation in melting point across a period reflects the structure and bonding of the elements.
- Across Period 3 electrical conductivity increases to aluminium then decreases.
- The density of the elements in Period 3 increases to aluminium then shows a general decrease.
- Electronegativity increases across a period.

Learning outcomes

On completion of this section, you should be able to:

- explain the variation in atomic and ionic radius of Period 3 elements
- describe the reactivity of Period 3 elements with oxygen, chlorine and water
- predict the type of bonding in the chlorides and oxides of Period 3 elements by reference to differences in ionic radius and electronegativity.

Did you know?

The term 'atomic radius' depends on the type of bond formed: metallic, covalent or van der Waals. For example, for sodium:

covalent radius = 0.134 nm
metallic radius = 0.191 nm
van der Waals radius = 0.230 nm

Covalent radii are generally used when comparing atom sizes across a period.

Patterns in atomic and ionic radii in Period 3

The **covalent radius** is often used as a measure of the size of an atom. This is half the distance between the nuclei of two atoms of the same type.

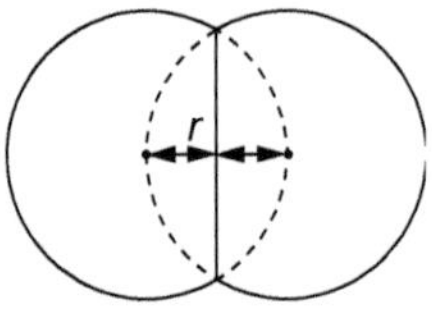

Figure 11.2.1 *The covalent radius of an atom, r, is half the distance between the two nuclei*

The atomic radii decrease across Period 3. As we move across the period, each successive element has one more proton in its nucleus. But the added electron goes into the same quantum shell (the third electron shell). So there is no additional shielding of outer shell electrons by the inner shells. The increase in nuclear charge from Na to Cl pulls the electrons closer to the nucleus. So the size of the atoms decreases across the period.

atom	Na	Mg	Al	Si	P	S	Cl
size of atom and nuclear charge	11+	12+	13+	14+	15+	16+	17+
Atomic (covalent) radius (nm)	0.156	0.136	0.125	0.117	0.110	0.104	0.099

Figure 11.2.2 *The atomic radii of the elements from sodium to chlorine*

- Metal ions are formed by losing electrons from their outer electron shell. So their ions are smaller than their atoms.
- Non-metal ions are formed by gaining electrons in their outer electron shell. There is more repulsion between the electrons than there is in the atom. This makes the negative ion larger than the atom from which it is derived.

The ionic radii decrease from Na^+ to Si^{4+}. As we move across the period, each successive ion has one more proton in its nucleus. But electronic structure is the same. The increase in nuclear charge from Na^+ to Si^{4+} pulls the electrons closer to the nucleus. So the size of the ions decreases. A similar explanation (higher nuclear charge and same electronic structure) accounts for the decrease in ionic radius from P^{3-} to Cl^-.

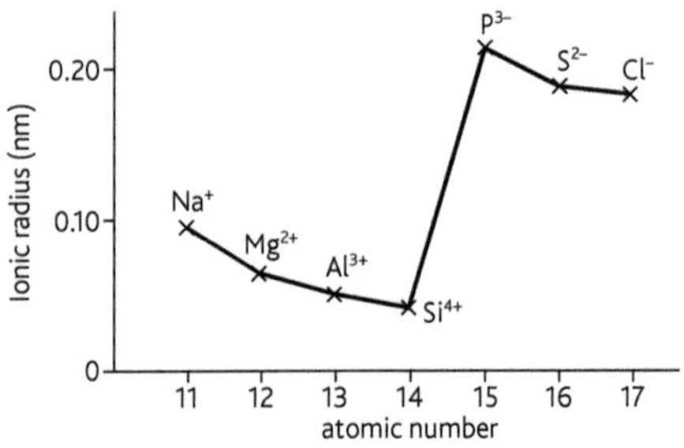

Figure 11.2.3 *The ionic radii from Na^+ to Cl^-*

The reaction of Period 3 elements with water

- Sodium reacts vigorously:

$$2Na(s) + 2H_2O(l) \rightarrow 2NaOH(aq) + H_2(g)$$

- Magnesium reacts very slowly with cold water but it reacts readily with steam to form magnesium oxide:

$$Mg(s) + H_2O(g) \rightarrow MgO(s) + H_2(g)$$

- Aluminium does not react with hot or cold water but it reacts with steam:

$$2Al(s) + 3H_2O(g) \rightarrow Al_2O_3(s) + 3H_2(g)$$

- Silicon, phosphorus and sulphur do not react. They are insoluble in water.
- Chlorine dissolves slightly in water and then reacts to form a mixture of hydrochloric and chloric(I) acid :

$$Cl_2(g) + H_2O(l) \rightleftharpoons 2H^+(aq) + Cl^-(aq) + ClO^-(aq)$$

The reaction of Period 3 elements with oxygen

- Sodium reacts vigorously when heated to form sodium oxide, Na_2O (although the final stable product is sodium peroxide, Na_2O_2):

$$2Na(s) + O_2(g) \rightarrow Na_2O_2(s)$$

- Magnesium and aluminium react vigorously with oxygen to form oxides:

$$2Mg(s) + O_2(g) \rightarrow MgO(s) \text{ and } 4Al(s) + 3O_2(g) \rightarrow 2Al_2O_3(s)$$

- Silicon reacts slowly with oxygen: $Si(s) + O_2(g) \rightarrow SiO_2(s)$
- Phosphorus reacts vigorously with oxygen to from phosphorus(V) oxide:

$$4P(s) + 5O_2(g) \rightarrow 2P_2O_5(s)$$

- Sulphur burns steadily in oxygen: $S(s) + O_2(g) \rightarrow SO_2(g)$
- Chlorine and argon do not combine directly with oxygen.

Did you know?

The bleaching action of moist chlorine is due to the formation of chloric(I) acid.

The reaction of Period 3 elements with chlorine

Apart from argon (and chlorine itself) all Period 3 elements react with chlorine.

Sodium, magnesium and aluminium react vigorously to form NaCl, $MgCl_2$ and $AlCl_3$ respectively, which are ionic solids. Silicon, phosphorus and sulphur react more slowly. For example:

with sodium:	$2Na(s) + Cl_2(g) \rightarrow 2NaCl(s)$
with silicon:	$Si(s) + 2Cl_2(g) \rightarrow SiCl_4(l)$
with phosphorus:	$2P(s) + 5Cl_2(g) \rightarrow 2PCl_5(s)$
with sulphur:	$2S(s) + Cl_2(g) \rightarrow S_2Cl_2(l)$

Key points

- The main influences on **atomic radius** in Period 3 are the size of the nuclear charge and the distance of the outer shell electrons from the nucleus.
- Across Period 3, the reactivity of the elements with oxygen, chlorine and water tends to decrease.
- Across Period 3, the chlorides and oxides change from ionic to covalent compounds.

11.3 Properties of Period 3 oxides and chlorides

Learning outcomes

On completion of this section, you should be able to:

- describe the variation in oxidation states of the chlorides and oxides of Period 3 elements
- describe the bonding in chlorides and oxides of Period 3 elements in terms of electronegativity and ionic radius and charge
- explain the trend in acid–base behaviour of oxides and hydroxides across Period 3.

Oxidation states

The oxides of elements in Period 3 all exist in positive oxidation states because oxygen is more electronegative than any of these elements. The maximum oxidation state of each element in Period 3 in their oxides rises across the period. This corresponds to the ability of the elements to use all the electrons in their outermost electron shells in bonding. This leads to an expanded octet for oxides of P, S and Cl.

Oxide	Na_2O	MgO	Al_2O_3	SiO_2	P_4O_{10}	SO_3	Cl_2O_7
Max. oxidation state	+1	+2	+3	+4	+5	+6	+7

The non-metallic oxides can also form oxides in lower oxidation states e.g. SiO, P_2O_3, SO_2 and several lower oxides of chlorine (Cl_2O, ClO_2, Cl_2O_6).

The chlorides of Period 3 elements also exist in positive oxidation states because chlorine is more electronegative that the other elements in the period. The oxidation states of the chlorides show a similar pattern, rising to +5 in PCl_5. The maximum oxidation state for sulphur in its chlorides, however, is +2.

Chloride	NaCl	$MgCl_2$	$AlCl_3$	$SiCl_4$	PCl_5	SCl_2
Max. oxidation state	+1	+2	+3	+4	+5	+2

Some of the non-metallic elements can also form chlorides in lower oxidation states, e.g. PCl_3, S_2Cl_2.

Did you know?

Many ionic compounds have a degree of covalent character because of ion polarisation. This is when a small highly-charged cation distorts the electron cloud charge of a much larger anion. Ion polarisation is most likely to occur when **i** the cation has a high charge and is small, **ii** the anion has a high charge and is large.

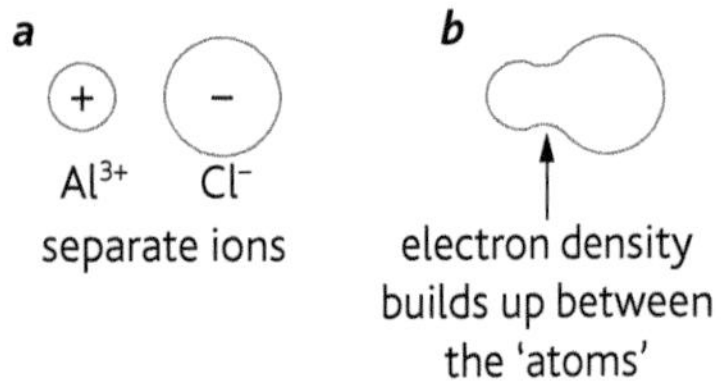

Figure 11.3.1 *The polarisation of a large Cl^- ion by a small, highly charged Al^{3+} ion*

Bonding in chlorides and oxides of Period 3

The structure and bonding of these compounds can be related the relative electronegativity of the atoms involved in bonding. The greater the difference in electronegativity, the more likely it is that the oxide or chloride will be ionic. The electronegativity difference between oxygen and Na, Mg, or Al is so great that one, two or three electrons respectively are transferred from the metal atom to the non-metal atom. These oxides are therefore ionic. The other Period 3 oxides are covalently bonded. Silicon dioxide has a giant covalent structure (see Section 2.6). Although in theory, an Si^{4+} ion can exist, the fourth ionisation energy is so large that the lattice energy cannot compensate for it to make ionic SiO_2 stable. Oxides of P, S and Cl have a simple molecular structure.

Electronegativity differences can also be used to explain the structure of Period 3 chlorides. Sodium chloride and magnesium chloride are ionic. But anhydrous aluminium chloride is a covalently-bonded molecule, Al_2Cl_6. The ionic radius of aluminium is very small and the Al^{3+} ion is highly charged. This high positive charge density tends to pull the electrons in the larger chloride ion towards it to such an extent that electron pairs are shared between the Al and Cl atoms. This is called **ion polarisation**.

Chlorides of Si, P and S have a simple molecular structure.

Reactions of Period 3 oxides and hydroxides

The oxides of sodium and magnesium react with water to form hydroxides:

$$Na_2O(s) + H_2O(l) \rightarrow 2NaOH(aq)$$

Magnesium hydroxide is less alkaline because it is less soluble than sodium hydroxide. Oxides and hydroxides of sodium and magnesium are basic. They react with acids to form a salt and water. e.g.

$$MgO(s) + 2HCl(aq) \rightarrow MgCl_2(aq) + H_2O(l)$$

Aluminium oxide does not dissolve in water but reacts with acids and alkalis:

$$Al_2O_3(s) + 3H_2SO_4(aq) \rightarrow Al_2(SO_4)_3 + 3H_2O$$

$$Al_2O_3(s) + 2NaOH(aq) + 3H_2O(l) \rightarrow \underset{\text{sodium aluminate}}{2NaAl(OH)_4(aq)}$$

A substance which acts both as an acid and a base is said to be **amphoteric**.

Silicon dioxide is insoluble in water but like Al_2O_3 reacts with hot alkali:

$$SiO_2(s) + 2NaOH(aq) \rightarrow Na_2SiO_3(aq) + H_2O(l)$$

SiO_2 is an acidic oxide. It does not react with acids.

Oxides of P, S and Cl are all acidic oxides. They react with water to form acidic solutions and with alkalis to form salts, e.g.

$$SO_2(g) + H_2O(l) \rightarrow H_2SO_3(aq)$$

$$P_4O_6(g) + 6H_2O(l) \rightarrow 4H_3PO_3(aq)$$

Reactions of the chlorides of Period 3

Chlorides of sodium and magnesium and *hydrated* aluminium chloride dissolve in water because they are ionic.

The chlorides of silicon and phosphorus are hydrolysed by water to form acidic solutions. Fumes of hydrogen chloride are also released.

$$SiCl_4(l) + 2H_2O(l) \rightarrow SiO_2(s) + 4HCl(g)$$

$$PCl_3(s) + 3H_2O(l) \rightarrow H_3PO_3(aq) + 3HCl(g)$$

$$PCl_5(s) + 4H_2O(l) \rightarrow H_3PO_4(aq) + 5HCl(g)$$

The hydrolysis of S_2Cl_2 is complex, several sulphur compounds including thionic acid being formed.

Did you know?

Anhydrous aluminium chloride has a simple molecular structure. It behaves differently to hydrated aluminium chloride. Aluminium chloride solution is hydrolysed by water to form an acidic solution:

$$[Al(H_2O)_6]^{3+} \rightarrow [Al(H_2O)_5OH]^{2+} + H^+$$

Key points

- The oxidation states of oxides and chlorides from Na to Al are fixed. The rest, apart from Ar, have several oxidation states.
- Chlorides of Na and Mg dissolve in water to form ionic solutions. The other chlorides are hydrolysed to form acidic solutions.
- Across Period 3, the oxides change from basic to amphoteric (Al) to acidic.
- Sodium and magnesium oxides and hydroxides are basic. Aluminium oxide is amphoteric. The oxides of Si, P, S and Cl are acidic.

12 The chemistry of Groups II, IV and VII

12.1 Properties of Group II elements

Learning outcomes

On completion of this section, you should be able to:

- explain the variation in properties of Group II elements in terms of structure and bonding
- describe the reactions of Group II elements with oxygen, water and dilute acids
- state some uses of MgO, CaO, $Ca(OH)_2$ and $CaCO_3$.

Physical properties

Elements in the periodic table can be put into blocks according to the type of orbitals, s, p, d or f in their outer shell. Group II elements are in the s block in the Periodic Table. They have two s-type orbitals in their outer shell.

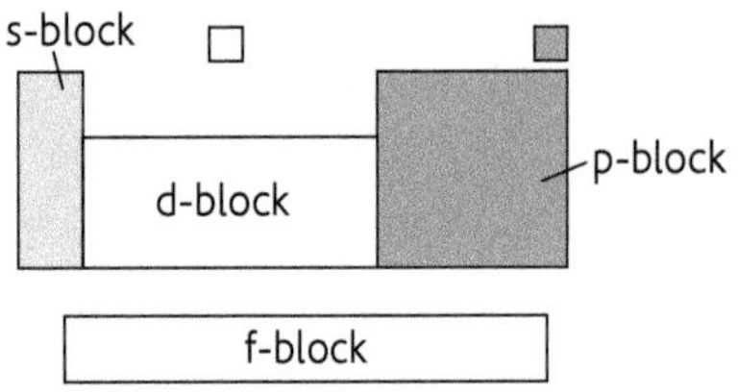

Figure 12.1.1 *The s, p, d and f blocks in the Periodic Table*

The table shows some physical properties of these elements.

Element	Atomic number	Metallic radius/nm	Ionic radius/nm	Density/ $g\,cm^{-3}$	Melting point/°C	Ionisation energies/$kJ\,mol^{-1}$	
						First	Second
beryllium, Be	4	0.122	0.031	1.85	1278	900	1760
magnesium, Mg	12	0.160	0.065	1.74	694	763	1450
calcium, Ca	20	0.197	0.099	1.55	839	590	1150
strontium, Sr	38	0.215	0.113	2.60	769	548	1060
barium, Ba	56	0.224	0.135	3.51	725	502	966

Did you know?

The metallic radius is half the distance between two neighbouring nuclei in a metallic giant structure. For s block elements, the metallic radius is greater than the covalent radius. The diagram shows the metallic radius, *r*.

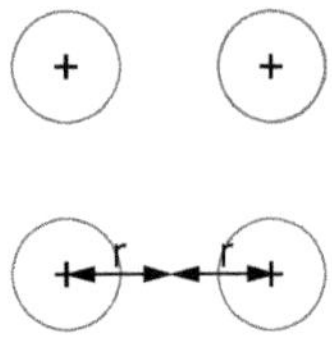

Metallic and ionic radius

The metallic and ionic radii both increase down the group as the number of filled electron shells increases. The ionic radius is much smaller than the **metallic radius** because when forming ions the electrons in the outer shell have been lost.

Density

The density depends on the mass of the atoms and the way they are packed in the lattice. From calcium to barium there is an increase in density. Beryllium and magnesium have higher densities than calcium because they are packed more efficiently in the lattice with less wasted space.

Melting points

The melting points are high, as expected of a giant metallic structure. As we go down the group there is a general decrease in melting points. Magnesium spoils this trend because it has a different lattice structure to Ca, Sr and Ba. Going down the group the electrons which contribute to the sea of delocalised electrons are further away from the positive nuclei. So there is less attraction between these electrons and the nuclei and it takes less energy to overcome these attractive forces. The boiling points follow a similar trend.

First and second ionisation energies

As we go down the group, the outer electrons are:

- further away from the nucleus
- screened more effectively by the greater number of inner electron shells.

These factors outweigh the effect of the increased nuclear charge and so it is easier to remove these electrons as we go down the group.

Some reactions of Group II metals

When Group II meals react, they lose the two electrons from their outer shell. So they act as reducing agents. From magnesium to barium, they generally exhibit similar reactions. They are more reactive down the group because it is easier to lose the outer electrons. Beryllium has chemical and some physical properties which are more like those of aluminium than the other Group II elements. So its reactions are not considered here.

Reaction with oxygen

This is a redox reaction. The reactivity with oxygen increases down the group.

$$2Mg(s) + O_2(g) \rightarrow 2MgO(s)$$

The oxides are basic in character, reacting with water to form hydroxides:

$$CaO(s) + H_2O(l) \rightarrow Ca(OH)_2(s)$$

Reaction with water

Magnesium reacts very slowly with cold water but more vigorously with steam to form magnesium oxide and hydrogen (see Section 11.2).

Calcium reacts vigorously with cold water to form calcium hydroxide:

$$Ca(s) + 2H_2O(l) \rightarrow Ca(OH)_2(s) + H_2(g)$$

A white precipitate is observed, but some $Ca(OH)_2$ dissolves to form a weakly alkaline solution. The other elements down the group react in a similar manner with increasing vigour. The resulting solutions get more alkaline down the group as the solubility of the hydroxides produced increases.

Reaction with acids

All members of the group react readily with acids to form hydrogen and the salt of the metal. The reactivity increases down the group:

$$Mg(s) + 2HCl(aq) \rightarrow MgCl_2(aq) + H_2(g)$$
$$Ba(s) + H_2SO_4(aq) \rightarrow BaSO_4(s) + H_2(g)$$

Uses of some compounds of Group II

- Calcium oxide: making cement and mortar; drying agent
- Calcium hydroxide: neutralising acidic soil; making bleaching powder; making limewater
- Calcium carbonate: limestone blocks for building; removing SiO_2 as slag in the blast furnace for the extraction of iron; making calcium oxide for cement
- Magnesium oxide: for lining furnaces

Key points

- Group II elements from Mg to Ba are metals which react with water to produce hydrogen and a metal hydroxide.
- The atomic radii of the metals increases down the group.
- Group II metals burn in air to form oxides which are more soluble down the group.
- The reactivity of Group II elements with oxygen, water and dilute hydrochloric acid increases down the group.
- Many compounds of Group II metals have important uses.

12.2 Group II carbonates, nitrates and sulphates

Learning outcomes

On completion of this section, you should be able to:

- explain the variation in thermal decomposition of Group II carbonates and nitrates
- explain the variation of the solubility of Group II sulphates
- know the test for sulphate ions using barium chloride

Thermal decomposition of carbonates

Group II carbonates from magnesium to barium, decompose on heating to form the oxide and carbon dioxide:

$$CaCO_3(s) \rightarrow CaO(s) + CO_2(g)$$

The temperatures at which these carbonates are 50% decomposed are shown below.

Carbonate	$MgCO_3$	$CaCO_3$	$SrCO_3$	$BaCO_3$
Temperature/K	813	1173	1553	1633

Going down Group II, the carbonates become more resistant to thermal decomposition. All Group II cations have the same charge (+2) but as we go down the group:

- The size of the metal cation increases.
- The smaller the cation, the better it is at distorting the electron cloud charge of the larger carbonate ion.
- Smaller ions are better polarisers of large ions (see Section 11.3).
- So Group II carbonates with smaller ions have a greater degree of covalence in the ionic bonding.
- The greater the degree of covalence, the less is the energy required to break a C=O bond in the carbonate ion.

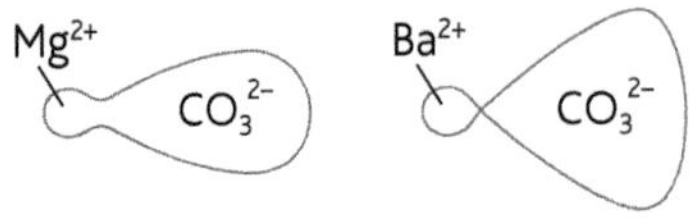

Figure 12.2.1 ***a** The small Mg^{2+} ion is a good polariser of the carbonate ion;* ***b** The larger Ba^{2+} ion is a poor polariser of the carbonate ion*

Thermal decomposition of nitrates

Group II nitrates from magnesium to barium, decompose on heating to form the oxide, nitrogen dioxide and oxygen:

$$2Mg(NO_3)_2(s) \rightarrow 2MgO(s) + 4NO_2(g) + O_2(g)$$

Going down Group II, the nitrates also become more resistant to thermal decomposition. The smaller the cation:

- The better it is at distorting the electron cloud charge of the larger nitrate ion.
- The greater the degree of covalence in the ionic bonding.
- The less energy required to break a particular N–O bond in the nitrate ion.

Solubility of Group II sulphates

The solubility of Group II sulphates in water decreases down the group.

Magnesium sulphate is very soluble. Barium sulphate is 'insoluble'. Solubility depends on the enthalpy change of solution, $\Delta H^{\ominus}_{sol}$. As a rough guide, the more endothermic the value of $\Delta H^{\ominus}_{sol}$, the less soluble is the salt. The enthalpy change of solution depends on the relative values of the lattice energy, $\Delta H^{\ominus}_{latt}$ and the enthalpy changes of hydration of the aqueous ions, $\Delta H^{\ominus}_{hydr}$ (see Section 6.3).

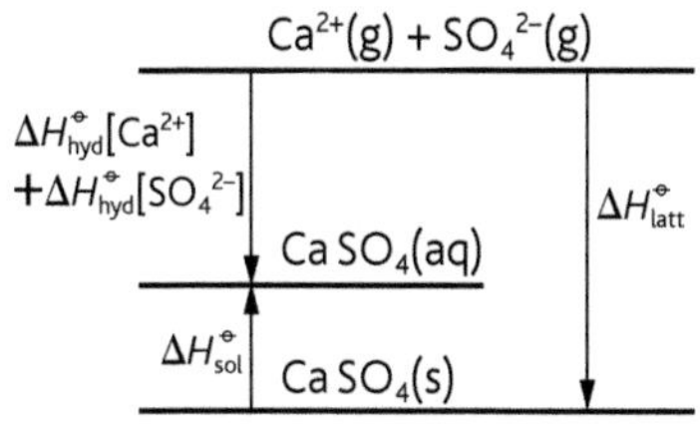

***Figure 12.2.2** An enthalpy cycle for calculating $\Delta H^{\ominus}_{sol}$*

- Lattice energy decreases down the group in the order:

$$MgSO_4 > CaSO_4 > SrSO_4 > BaSO_4$$

This decrease is fairly small because it is determined more by the size of the large sulphate ion than by the cation.

- The enthalpy change of hydration also increases in the same order. But this decrease is relatively larger because it is determined more by the size of the cations than the anion.
- By Hess's law: $\Delta H^{\ominus}_{sol} = \Delta H^{\ominus}_{hyd} - \Delta H^{\ominus}_{latt}$
- So $\Delta H^{\ominus}_{hyd} - \Delta H^{\ominus}_{latt}$ gets more endothermic down the group (the value of $\Delta H^{\ominus}$ gets more positive). So the solubility decreases.

Testing for sulphate ions

The relative insolubility of barium sulphate is used as the basis for the following test to detect sulphate ions:

- Acidify the solution to be tested with nitric acid or hydrochloric acid. This removes contaminating carbonate ions, which react with the acid to form carbon dioxide. If these were not removed insoluble white precipitate of barium carbonate would interfere with the test.
- Add aqueous barium chloride or aqueous barium nitrate.
- If sulphate ions are present a white precipitate of barium sulphate is formed:

$$Ba^{2+}(aq) + SO_4^{2-}(aq) \rightarrow BaSO_4(s)$$

Key points

- Group II carbonates and nitrates are more resistant to thermal decomposition down the group. This is because of the decreasing degree of polarisation of the large CO_3^{2-} or NO_3^- ion by the increasingly large metal cation.
- The decreasing solubility of the Group II sulphates down the group can be explained in terms of the relative values of their enthalpy changes of hydration and lattice energies.
- When barium chloride is added to a solution containing a sulphate, a white precipitate is formed.

12.3 Group IV elements and their tetrachlorides

Learning outcomes

On completion of this section, you should be able to:

- explain the variation in physical properties of the Group IV elements in terms of their structure and bonding
- describe the bonding of the Group IV tetrahalides
- describe the reaction of Group IV tetrahalides with water.

Did you know?

Tin exists in several forms, depending on the temperature. At low temperatures, tin exists as a powdery grey form which has a structure similar to diamond. In countries which have cold climates, articles made of tin may become damaged by the cold if they are left outside for too long.

Some physical properties of Group IV elements

The Group IV elements are in the p block of the Periodic Table. They all have four electrons in their outer principal quantum shell. They are all solids at room temperature. Some of their physical properties are shown in the table.

Element	Structure	Melting point/°C	Electrical conductivity
carbon (diamond), C	non-metal	3550	non-conductor
silicon, Si	metalloid	1410	semi-conductor
germanium, Ge	metalloid	937	semi-conductor
tin, Sn	metal	232	conductor
lead, Pb	metal	327	conductor

Trends in melting points

- Si and Ge have giant covalent structures similar to that of C(diamond) (see Section 2.6). They all have very high melting points due to the fact that a lot of energy is needed to break the many covalent bonds in the structure.
- The decrease in melting point from C(diamond) to Ge reflects the decreasing amount of energy needed to break the bonds:

$$E(\text{C}-\text{C}) = +350\,\text{kJ mol}^{-1} \quad E(\text{Si}-\text{Si}) = +222\,\text{kJ mol}^{-1}$$
$$E(\text{Ge}-\text{Ge}) = +188\,\text{kJ mol}^{-1}$$

- Sn and Pb have relatively low melting points compared with most other metals. The strength of metallic bonding decreases as the size of the ions increases (see Section 2.4). Tin and lead have fairly large ions, so have relatively low melting points.

Electrical conductivity

- There is a general trend in electrical conductivity down the group.
- Carbon in the form of diamond does not conduct electricity because all its outer shell electrons are used in covalent bonding. Graphite does conduct because some of its electrons are delocalised and so are free to move (see Section 2.6).
- Si and Ge conduct electricity to a very small extent. They do not have delocalised electrons. Some of their electrons can move out of position by 'jumping' to different places in the lattice. Their conductivity increases with increase in temperature (as opposed to metals, where conductivity decreases with increase in temperature). Si and Ge are described are **semi-conductors**. Their structure is described as **metalloid**.

- Sn and Pb are electrical conductors because their outer shell electrons are delocalised. Sn is a slightly better conductor than Pb. In lead there is a greater force of attraction between its ions and the delocalised electrons compared with Sn. This is because the effect of the increased nuclear charge from Sn to Pb is greater than the effect of shielding and increase in atomic radius. This results in lead having higher first and second ionisation energies than Sn. This is opposite to the general trend of ionisation energies down a group (see Section 1.5).

The Group IV tetrachlorides

All the elements in Group IV form tetrachlorides, XCl_4. The tetrachlorides are all simple covalent molecules with tetrahedral structure.

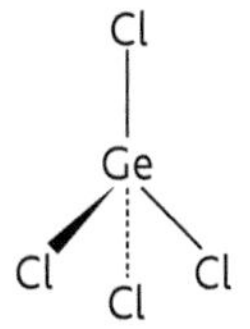

Figure 12.3.1 *The tetrahedral structure of germanium(IV) chloride*

All the tetrachlorides are volatile liquids – they have low boiling points. This is because they are simple covalent molecules with only weak van der Waals forces between their molecules. They are non-polar because the dipoles cancel. Although CCl_4 and $SiCl_4$ have lower boiling points than $GeCl_4$, $SnCl_4$, and $PbCl_4$, there is no simple pattern in the boiling points:

$$CCl_4 = 76\,°C \qquad SiCl_4 = 57\,°C \qquad GeCl_4 = 87\,°C$$
$$SnCl_4 = 114\,°C \qquad PbCl_4 = 105\,°C$$

Reactivity of the Group IV tetrahalides with water

All the tetrahalides, with the exception of CCl_4, are readily hydrolysed by water to the oxide in the +4 oxidation state. Acidic fumes of hydrogen chloride are also produced, e.g.

$$SiCl_4(l) + 2H_2O(l) \rightarrow SiO_2(s) + 4HCl(g)$$

The hydrolysis of the heavier tetrahalides is reversible. For example:

$$GeCl_4(l) + 2H_2O(l) \rightleftharpoons GeO_2(s) + 4HCl(g)$$

The ease of hydrolysis increases down the group from $SiCl_4$ to $PbCl_4$ as the metallic nature of the Group IV atom increases. When lead(IV) chloride is hydrolysed, a little decomposition of the lead(IV) chloride to lead(II) chloride also occurs.

Key points

- C, Si and Ge have giant covalent structures; Sn and Pb are metals.
- Down Group IV, the elements show a general decrease in melting point and a general increase in metallic character and electrical conductivity.
- Group IV elements form covalently bonded tetrachlorides.
- The Group IV tetrahalides are hydrolysed rapidly by water with the exception of carbon tetrachloride.

12.4 Some properties of Group IV oxides

Learning outcomes

On completion of this section, you should be able to:

- describe the trends in bonding and thermal stability of the Group IV oxides
- describe the trends in acid–base character of Group IV oxides in oxidation states +2 and +4.

The Group IV oxides: structure and stability

The oxides of the Group IV elements exist in the +2 and +4 oxidation states. The table shows the type of structure of these oxides.

+2 oxidation state		**+4 oxidation state**	
CO	simple molecular	CO_2	molecular
SiO	simple molecular	SiO_2	giant covalent
GeO	ionic with some covalent character	GeO_2	giant covalent
SnO	giant ionic	SnO_2	giant covalent with ionic character
PbO	giant ionic	PbO_2	giant covalent with ionic character

Down each series of oxides, the structures get more ionic as the metallic character of the Group IV elements increases. SnO_2 and PbO_2 have some degree of covalent character in their ionic structure. All the oxides, apart from CO and CO_2, are solids with giant structures and high melting points. CO and CO_2 are gases at room temperature, because they have a simple molecular structure with only weak forces between their molecules in the solid and liquid states.

- Carbon monoxide, CO has a strong triple bond and does not decompose on heating.
- Germanium(II) oxide disproportionates on heating:

$$2GeO(s) \rightarrow GeO_2(s) + Ge(s)$$

Oxidation numbers +2 +4 0

- Tin(II) oxide, SnO, and lead(II) oxide, PbO, do not decompose on heating in the absence of air. They are readily oxidised to higher oxides in the presence of oxygen.
- Carbon dioxide, CO_2 has strong double bonds and does not decompose on heating.
- The oxides in the +4 oxidation state tend to decrease in stability down the group. Only lead(IV) oxide, PbO_2, however, undergoes significant thermal decomposition:

$$\underset{\text{lead(IV) oxide}}{PbO_2(s)} \rightarrow \underset{\text{lead(II) oxide}}{PbO(s)} + O_2(g)$$

Acid–base properties of the oxides

The table below summarises the acid–base properties of the oxides in the +2 and +4 oxidation states.

+ 2 oxidation state		**+ 4 oxidation state**	
CO	very weakly acidic	CO_2	acidic
SiO	–	SiO_2	very weakly acidic
GeO	amphoteric	GeO_2	amphoteric
SnO	amphoteric	SnO_2	amphoteric
PbO	amphoteric	PbO_2	amphoteric

- The oxides in the +2 oxidation state are less acidic (more basic) than the corresponding oxides in the +4 oxidation state. For example:

  CO only reacts with hot concentrated sodium hydroxide, but CO_2 forms an acidic solution in water:

  $$CO_2(g) + H_2O(l) \rightleftharpoons HCO_3^-(aq) + H^+(aq)$$

  CO_2 also reacts with dilute alkalis:

  $$CO_2(g) + 2NaOH(aq) \rightarrow Na_2CO_3(aq) + H_2O(l)$$

- The oxides in both oxidation states become more basic down the group as the metallic character of the Group IV atom increases. For example:

  Carbon dioxide reacts with aqueous alkalis (hydroxide ions) but silicon dioxide only reacts with hot concentrated alkali:

  $$SiO_2(s) + 2NaOH(aq) \rightarrow \underset{\text{sodium silicate}}{Na_2SiO_3(aq)} + H_2O(l)$$

Reactions of the amphoteric oxides of Group IV

Oxidation state +2

- GeO, SnO and PbO are all amphoteric, but have increasingly basic character in the order GeO < SnO < PbO.
- They all react with acids to form a salt with oxidation state +2. For example:

  $$SnO(s) + 2HCl(aq) \rightarrow SnCl_2(aq) + H_2O(l)$$

- They all react with alkalis to form oxo ions with oxidation state +2. For example:

  $$GeO(s) + 2OH^-(aq) \rightarrow \underset{\text{germanate(II) ion}}{GeO_2^{2-}(aq)} + H_2O(l)$$
  $$SnO(s) + 2OH^-(aq) \rightarrow \underset{\text{stannate(II) ion}}{SnO_2^{2-}(aq)} + H_2O(l)$$
  $$PbO(s) + 2OH^-(aq) \rightarrow \underset{\text{plumbate(II) ion}}{PbO_2^{2-}(aq)} + H_2O(l)$$

Oxidation state +4

- GeO_2, SnO_2 and PbO_2 are all amphoteric.
- They all react with acids to form a salt with oxidation state +4. For example:

  $$SnO_2(s) + 4HCl(aq) \rightarrow SnCl_4(aq) + 2H_2O(l)$$

  SnO_2 reacts with dilute acid but PbO_2 will only react with concentrated acid, oxidising it to chlorine (see Section 12.5).

- They all react with alkalis to form oxo ions with oxidation state +4. For example:

  $$GeO_2(s) + 2OH^-(aq) \rightarrow \underset{\text{germanate(IV) ion}}{GeO_3^{2-}(aq)} + H_2O(l)$$
  $$SnO_2(s) + 2OH^-(aq) \rightarrow \underset{\text{stannate(IV) ion}}{SnO_3^{2-}(aq)} + H_2O(l)$$
  $$PbO_2(s) + 2OH^-(aq) \rightarrow \underset{\text{plumbate(IV) ion}}{PbO_3^{2-}(aq)} + H_2O(l)$$

Key points

- The relative stability of Group IV oxides in oxidation state +4 decreases down the group.
- The acidic character of the Group IV oxides in oxidation state +2 and +4 decreases down the group and the basic character increases.
- Group IV oxides in oxidation state +2 are more basic than the corresponding oxides in oxidation state +4.

12.5 Relative stability of Group IV cations and oxides

Learning outcomes

On completion of this section, you should be able to:

- describe the relative stabilities of the oxides and aqueous cations of the Group IV elements in oxidation states +2 and +4 with reference to $E^\ominus$ values
- describe some uses of ceramics based on silicon(IV) oxide.

Stability of +2 and +4 oxidation states

Going down Group IV, the oxides in the +2 oxidation state become more stable and the oxides of the +4 oxidation state become less stable with respect to the +2 state.

	oxidation state +2	oxidation state +4	
decreasing stability and better reducing agent ↑	CO	CO_2	↓ decreasing stability and better oxidising agent
	SiO	SiO_2	
	Ge	GeO_2	
	SnO	SnO_2	
	PbO	PbO_2	

- Carbon monoxide, CO, is a good reducing agent. It loses electrons readily, e.g. at a high temperature it reduces iron(III) oxide to iron. It is oxidised to carbon dioxide which is more stable.

$$Fe_2O_3(s) + 3CO(g) \rightarrow 2Fe(l) + 3CO_2(g)$$

Oxidation numbers: +3 +2 0 +4

- Carbon dioxide is a poor reducing agent. It loses electrons with great difficulty.
- Lead(IV) oxide, PbO_2, is a good oxidising agent. It accepts electrons readily, e.g. it oxidises hydrogen chloride to chlorine:

$$PbO_2(s) + 4HCl(aq) \rightarrow PbCl_2(aq) + Cl_2(g) + 2H_2O(l)$$

Oxidation numbers: +4 −1 +2 0

- Lead(II) oxide is a poorer oxidising agent in comparison with lead(IV) oxide.

Reactions of some Group(IV) cations

In Section 10.4, we showed how $E^\ominus$ values can be used to predict the feasibility of a reaction. We can apply these ideas to the aqueous cations of Group IV metals in oxidation states +2 and +4. See Figure 12.5.1.

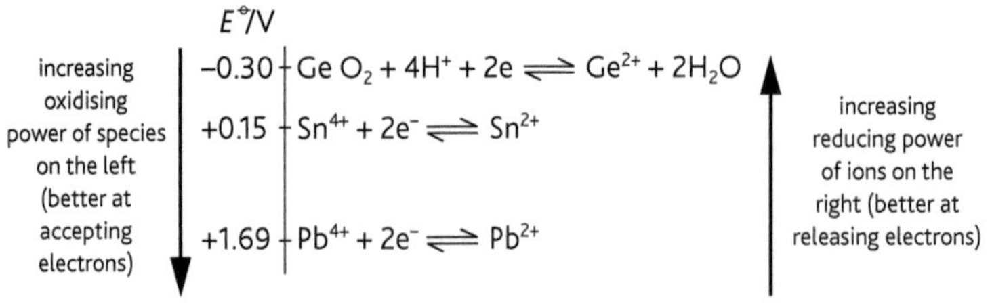

Figure 12.5.1 *The reducing and oxidising powers of some Group IV ions*

Worked examples 1 to 3 together with Figure 12.5.2 show how we can compare the ability of Pb^{4+} ions and Sn^{4+} ions to oxidise other species.

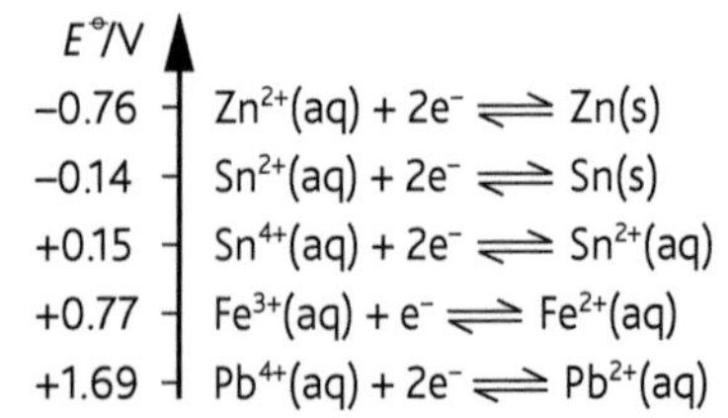

Figure 12.5.2 *An electrode potential diagram for determining the feasibility of some reactions involving Group IV cations*

Did you know?

Tables of $E^\ominus$ values often have tin and lead ions written in the form Pb^{4+} and Sn^{4+}. Many tin and lead compounds in the +4 oxidation state are insoluble in water. So they have to be dissolved in concentrated acids or alkalis to get these ions into solution.

Worked example 1

Will Pb^{4+} ions oxidise an aqueous solution containing Fe^{2+} ions to aqueous Fe^{3+} ions?

- According to the $E^{\ominus}$ values, Pb^{4+} ions are better at accepting electrons than Fe^{3+} ions. Pb^{4+} ions are better oxidising agents.
- According to the $E^{\ominus}$ values, Fe^{2+} ions are better at releasing electrons than Pb^{2+} ions. Fe^{2+} ions are better reducing agents.
- The reaction with the more negative $E^{\ominus}$ value goes in the reverse direction, i.e. $Fe^{2+}(aq) \rightarrow Fe^{3+}(aq)$
- The overall reaction is: $Pb^{4+}(aq) + 2Fe^{2+}(aq) \rightarrow Pb^{2+}(aq) + 2Fe^{3+}(aq)$

Worked example 2

Will Sn^{4+} ions oxidise an aqueous solution containing Fe^{2+} ions to aqueous Fe^{3+} ions?

- According to the $E^{\ominus}$ values, Fe^{3+} ions are better at accepting electrons than Sn^{4+} ions. Fe^{3+} ions are better oxidising agents.
- According to the $E^{\ominus}$ values, Sn^{2+} ions are better at releasing electrons than Fe^{2+} ions. Sn^{2+} ions are better reducing agents.
- The reaction with the more negative $E^{\ominus}$ value goes in the reverse direction, i.e. $Sn^{2+}(aq) \rightarrow Sn^{4+}(aq)$
- So no reaction occurs.

Worked example 3

Will Zn reduce Sn^{4+} ions?

- According to the $E^{\ominus}$ values, Zn are better at releasing electrons than Sn^{4+} ions. Zn is a better reducing agent.
- The reaction with the more negative $E^{\ominus}$ value goes in the reverse direction, i.e. $Zn(s) \rightarrow Zn^{2+}(aq)$
- The initial reaction is: $Zn(s) + Sn^{4+}(aq) \rightarrow Zn^{2+}(aq) + Sn^{2+}(aq)$
- But according to $E^{\ominus}$ values and the anticlockwise rule, Sn^{2+} can be reduced further by Zn to tin ($E^{\ominus}$– 0.14 V).
- The overall reaction is: $2Zn(s) + Sn^{4+}(aq) \rightarrow 2Zn^{2+}(aq) + Sn(s)$

Uses of ceramics based on silicon dioxide

- **Furnace linings:** SiO_2 is a good thermal insulator and has a very high melting point due to the many strong covalent bonds.
- **Abrasives:** SiO_2 is hard and has a high melting point (heat is given out on grinding).
- **Manufacture of glass and porcelain:** SiO_2 is relatively unreactive and is easily moulded. The high melting point is useful for ovenware.

Key points

- The relative stabilities of the oxides and aqueous cations of the Group IV elements can be explained with reference to $E^{\ominus}$ values.
- Down Group IV oxides, the +4 oxidation state becomes a better oxidising agent with respect to the +2 oxidation state.
- Ceramics based on silicon(IV) oxide are used for furnace linings and abrasives.

Exam tips

An easy way to determine if a reaction occurs or not is shown below. For example: will $Pb^{4+}(aq)$ ions react with $Fe^{2+}(aq)$ ions?

- Write down the two half equations involved making sure that you reverse the sign of the equation showing electron loss (in this case, equation **ii**).
  - i $Pb^{4+}(aq) + 2e^- \rightleftharpoons Pb^{2+}(aq)$ $E^{\ominus} = +1.69V$
  - ii $Fe^{2+}(aq) \rightleftharpoons Fe^{3+}(aq) + e^-$ $E^{\ominus} = -0.77V$
- Add the two voltages

  $+1.69 - 0.77 = +0.92V$
- If the value of the voltage is positive, the reaction will occur.

  So $Pb^{4+}(aq)$ reacts with $Fe^{2+}(aq)$.

12.6 Group VII: the halogens

Learning outcomes

On completion of this section, you should be able to:

- explain the variations in the physical properties of the halogens in terms of structure and bonding
- explain the relative reactivity of the halogens as oxidising agents
- describe the use of bromine water.

Physical properties

The elements of Group VII (the **halogens**) are in the p block of the Periodic Table. All have seven electrons in their outer principal quantum shell. They are all non-metals which exist as **diatomic** molecules (molecules having two atoms).

Halogen	State at 20 °C	Density/g cm^{-3}	Atomic radius/nm	Melting point/°C
fluorine, F_2	gas	1.51 (at 85 K)	0.072	−220
chlorine, Cl_2	gas	1.53 (at 238 K)	0.099	−101
bromine, Br_2	liquid	3.12 (at 293 K)	0.114	−7
iodine, I_2	solid	4.93 (at 293 K)	0.133	+114

Colour and solubility of the halogens

The colour gets darker down the group:

F_2: yellow; Cl_2: yellowish-green; Br_2: reddish-brown; I_2: grey-black.

Fluorine reacts with water. Chlorine reacts slightly with water. Solutions of Cl_2 and Br_2 are called 'chlorine water' and 'bromine water' respectively. Iodine is insoluble in water. Iodine solution is iodine dissolved in aqueous potassium iodide. This solution is brown. When iodine is dissolved in organic solvents such as cyclohexane, it appears purple. Iodine vapour is also purple.

Density and atomic radius

The atomic radius increases down the group as the number of electron shells increases and there is increased shielding of the outer shell electrons by the inner shells. The density increases down the group.

Melting points and boiling points

As the halogens get larger, the increasing number of electrons makes the van der Waals forces between the molecules stronger. So the melting points increase down the group as it takes more energy to break these intermolecular forces. The boiling points show a similar trend, F_2 and Cl_2 being gases at room temperature, bromine a volatile liquid and iodine a solid.

Chemical reactivity of the halogens

The halogens are less reactive going down the group. A more reactive halogen can displace a less reactive halogen from an aqueous halide solution. These are redox reactions.

Halogen	Halide ion		
	chloride, $Cl^-(aq)$	**bromide, $Br^-(aq)$**	**iodide, $I^-(aq)$**
Cl_2	–	turns orange	turns brown
Br_2	no reaction	–	turns from orange to brown
I_2	no reaction	no reaction	–

The reactions can be confirmed by adding cyclohexane to the solutions. The cyclohexane layer dissolves the halogen only, so any colour changes can be verified.

Aqueous chlorine displaces bromine from an aqueous solution of potassium bromide:

$$Cl_2(aq) + 2KBr(aq) \rightarrow 2KCl(aq) + Br_2(aq)$$

Ionic equation: $Cl_2(aq) + 2Br^-(aq) \rightarrow 2Cl^-(aq) + Br_2(aq)$

Oxidation numbers: 0   –1   –1   0

Iodine will not displace bromine from potassium bromide because the iodine is not as good an oxidising agent as bromine.

We can use $E^\ominus$ values to explain the relative reactivity of halogens as oxidising agents, e.g. does chlorine oxidise iodide ions to iodine?

- As $E^\ominus$ values get more positive (less negative), the halogens on the left become better oxidising agents. They accept electrons more readily.
- As $E^\ominus$ values get less positive (more negative), the halides on the right become better reducing agents. They release electrons more readily.
- Cl_2 accepts electrons more readily than I_2 and I^- ions release electrons more readily than Cl^- ions.
- The reaction with the more negative $E^\ominus$ value goes in the reverse direction, i.e.

$$I^-(aq) \rightarrow I_2(aq)$$

- So according to the anticlockwise rule, the reaction is feasible:

$$Cl_2(aq) + 2I^-(aq) \rightarrow 2Cl^-(aq) + I_2(aq)$$

$E^\ominus$/V	
+0.54	$I_2(aq) + 2e^- \rightleftharpoons 2I^-(aq)$
+1.07	$Br_2(aq) + 2e^- \rightleftharpoons 2Br^-(aq)$
+1.36	$Cl_2(aq) + 2e^- \rightleftharpoons 2Cl^-(aq)$
+2.87	$(F_2 + 2e^- \rightleftharpoons 2F^-)$

Figure 12.6.1 *Using $E^\ominus$ values to predict the feasibility of reaction between halogens and halides*

The use of bromine water to test for C=C bonds

Unsaturated compounds contain one or more C=C double bonds. Bromine water is used to test for double bonds in organic compounds. When an unsaturated compound is shaken with bromine water (aqueous bromine) the solution changes colour from orange (the colour of the bromine) to colourless:

$$\underset{\substack{\text{ethene}\\\text{(colourless)}}}{CH_2{=}CH_2} + \underset{\substack{\text{bromine}\\\text{(orange)}}}{Br_2} \rightarrow \underset{\substack{\text{1,2-dibromoethane}\\\text{(colourless)}}}{CH_2Br{-}CH_2Br}$$

Key points

- Halogens exist as diatomic molecules.
- Down the group, the halogens become less volatile and are darker in colour.
- The halogens are good oxidising agents, the oxidising ability decreasing down the group.
- Bromine water is used to test for unsaturated (C=C) bonds in carbon compounds.

Learning outcomes

On completion of this section, you should be able to:

- describe the redox reaction of iodine with sodium thiosulphate
- describe the reaction of the halogens with hydrogen
- explain the relative thermal stabilities of the Group VII hydrides.

Did you know?

Starch is a complex carbohydrate containing chains of glucose units, some of which may be arranged in a helical form. Iodine forms a weak complex with these helices. This complex is a dark blue-black colour. The complex is so weak that iodine can still react with species such as sodium thiosulphate. So when all the iodine has been used up, only the colourless starch remains.

Reaction of halogens with sodium thiosulphate

Sodium thiosulphate, $Na_2S_2O_3$, can be used to determine the concentration of iodine by titration (see Section 3.6) and to analyse samples of bleach. Figure 12.7.1 shows that this reaction is feasible because $S_2O_3^{2-}$ ions are better reducing agents than I^- ions and I_2 molecules are better oxidising agents than $S_4O_6^{2-}$ ions.

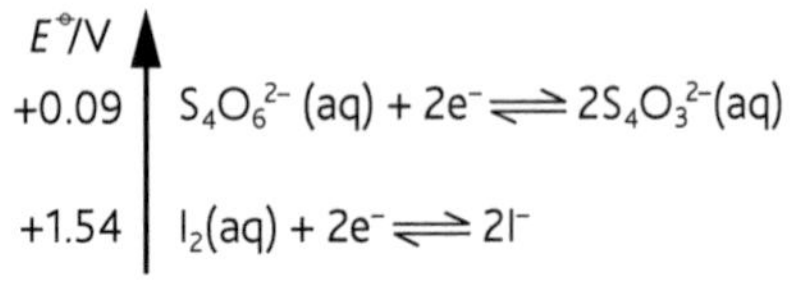

Figure 12.7.1

So the reaction is:

$$2Na_2S_2O_3(aq) + I_2(aq) \rightarrow Na_2S_4O_6(aq) + 2NaI(aq)$$

Bromine and chlorine are stronger oxidising agents than iodine and cause further oxidation of sodium thiosulphate to sulphate ions:

$$S_2O_3^{2-}(aq) + 4Cl_2(aq) + 5H_2O(l) \rightarrow 2SO_4^{2-}(aq) + 8Cl^-(aq) + 10H^+(aq)$$

The estimation of chlorine in bleaches

Commercial bleaches usually contain sodium chlorate(I), NaClO. This is commonly called 'sodium hypochlorite'. The chlorine in bleach can be determined by titration. The procedure is:

- Dilute the bleach by a known amount with distilled water.
- Add excess acidified potassium iodide solution to liberate iodine:

  $$NaClO(aq) + 2I^-(aq) + 2H^+(aq) \rightarrow I_2(aq) + NaCl(aq) + H_2O(l)$$
- Titrate the liberated iodine with sodium thiosulphate of a known concentration using starch indicator.
- The end point is when the blue-black colour of the starch–iodine indicator turns colourless.

Hydrogen halides

Hydrogen halides are formed when hydrogen combines directly with the halogens. These reactions are slower down the halogen group.

- Hydrogen fluoride, HF: The reaction is explosive even in cool conditions.

  $$H_2(g) + F_2(g) \rightarrow 2HF(g)$$
- Hydrogen chloride, HCl: The reaction is explosive in the presence of sunlight.

  $$H_2(g) + Cl_2(g) \rightarrow 2HCl(g)$$
- Hydrogen bromide, HBr: H_2 gas and Br_2 vapour react slowly on heating.
- Hydrogen iodide, HI: H_2 gas and I_2 vapour react slowly on heating in a closed container to form an equilibrium mixture:

  $$H_2(g) + I_2(g) \rightleftharpoons 2HI(g)$$

Hydrogen fluoride boils at 19.5 °C and so its state under normal laboratory conditions may be either liquid or gas. The other hydrogen halides are gases at room temperature. The much higher boiling point of HF compared with the other hydrogen halides is due to its extensive hydrogen bonding.

All hydrogen halides are acidic – they form acids when they dissolve in water e.g.

$$HCl(g) + H_2O(l) \rightarrow H_3O^+(aq) + Cl^-(aq)$$

or more simply: $$HCl(g) + aq \rightarrow H^+(aq) + Cl^-(aq)$$

The thermal stability of hydrogen halides

The thermal stability of the hydrogen halides decreases as the halogen atom increases in size.

- HF and HCl are not decomposed at temperatures of 1500 °C
- HBr decomposes slightly at about 450 °C to form hydrogen and bromine.
- HI decomposes rapidly at about 450 °C to form hydrogen and iodine:

$$2HI(g) \rightleftharpoons H_2(g) + I_2(g)$$

- When a hot platinum wire is placed in a tube of hydrogen iodide a purple vapour of iodine is first seen, which then turns directly to a grey-black solid on the side of the tube. The direct formation of a solid from a gas or vice versa omitting the liquid phase is called **sublimation**.

The ease of thermal decomposition of the hydrogen halides is related to the bond energies of the hydrogen–halogen bond:

Bond	Bond energy/kJ mol^{-1}
H–F	562
H–Cl	431
H–Br	366
H–I	299

The bond energy decreases in this way because:

- The larger the halogen atom, the greater is the distance between the hydrogen and halogen nuclei.
- So down the group there is a smaller attractive force between the nuclei and the bonding electrons.
- So going down the group, it takes less energy to break the carbon–halogen bond.

Key points

- The reaction of iodine with sodium thiosulphate can be explained by reference to $E^{\ominus}$ values.
- The halogens react with hydrogen to form hydrogen halides.
- The relative thermal stabilities of the hydrogen halides decrease down the group as the hydrogen–halogen bond strength decreases.

12.8 Reactions of halide ions

Learning outcomes

On completion of this section, you should be able to:

- describe the reactions of halide ions with aqueous silver nitrate followed by aqueous ammonia
- describe the reaction of halide ions with concentrated sulphuric acid
- describe the reaction of chlorine with cold and hot aqueous solutions of sodium hydroxide.

Exam tips

Remember that the redox reactions between halides and sulphuric acid involve the solid halides (not the aqueous solutions).

The reaction of halide ions with silver nitrate

When silver nitrate is added to an aqueous solution of Cl^-, Br^- or I^- ions, characteristically coloured precipitates of silver halides are formed. Fluoride ions cannot be tested for in this way because silver fluoride is soluble in aqueous solution and no precipitate is formed.

The procedure is:

- Add dilute nitric acid to the solution under test. This removes any soluble contaminating carbonates or hydroxides which may form insoluble silver compounds.
- Add a few drops of aqueous silver nitrate until a precipitate is seen.
- Record the colour of the precipitate.
- Add dilute aqueous ammonia to see if the precipitate redissolves.
- If the precipitate does not redissolve, add concentrated aqueous ammonia.

Halide ion	**Cl^-**	**Br^-**	**I^-**
Colour of precipitate	white	cream	pale yellow
Confirmatory test	dissolves in dilute $NH_3(aq)$	dissolves in concentrated $NH_3(aq)$	insoluble in concentrated $NH_3(aq)$

If left exposed to sunlight:

- The silver chloride precipitate turns grey rapidly.
- The silver bromide precipitate turns grey very slowly.
- The silver iodide precipitate does not turn grey.

In each case, the equations for the precipitation reaction are similar e.g.

$$AgNO_3(aq) + KBr(aq) \rightarrow AgBr(s) + KNO_3(aq)$$

Ionic equation: $Ag^+(aq) + Br^-(aq) \rightarrow AgBr(s)$

The precipitate of AgCl dissolves in excess ammonia because a **complex ion** is formed (See Section 14.3).

Reaction with concentrated sulphuric acid

The strength of halide ions as reducing agents follows the pattern $I^- > Br^- > Cl^-$ with iodide being the best reducing agent. The larger the halide ion, the easier it is to lose an electron from their outer shell because there is less force of attraction between the nucleus and the outer electrons.

Concentrated H_2SO_4 and solid NaCl

Chloride ions are not strong enough reducing agents to reduce the S in H_2SO_4. An acid–base reaction occurs:

$$NaCl(s) + H_2SO_4(l) \rightarrow NaHSO_4(s) + HCl(g)$$

Concentrated H_2SO_4 and solid NaBr

When bromide ions react with H_2SO_4 there is first an acid–base reaction:

$$NaBr(s) + H_2SO_4(l) \rightarrow NaHSO_4(s) + HBr(g)$$

Further reaction occurs because bromide ions are strong enough reducing agents to reduce S from the the +6 oxidation state in H_2SO_4 to the +4 oxidation state in sulphur dioxide, SO_2.

$$2HBr(g) + H_2SO_4(l) \rightarrow SO_2(g) + Br_2(l) + 2H_2O(l)$$

OxNo: −1 +6 +4 0

Concentrated H_2SO_4 and solid NaI

Hydrogen iodide is first formed in an acid–base reaction similar to the ones above. Iodide ions are better reducing agents than bromide ions. They are strong enough reducing agents to reduce S from the the +6 oxidation state in H_2SO_4 to the −2 oxidation state in hydrogen sulphide, H_2S. A mixture of products may be formed:

$$2HI(g) + H_2SO_4(l) \rightarrow SO_2(g) + I_2(l) + 2H_2O(l)$$

$$6HI(g) + H_2SO_4(l) \rightarrow S(s) + 3I_2(s) + 4H_2O(l)$$

OxNo: −1 +6 0 0

$$8HI(g) + H_2SO_4(l) \rightarrow H_2S(g) + 4I_2(s) + 4H_2O(l)$$

OxNo: −1 +6 −2 0

The reaction of chlorine with alkalis

With cold dilute alkali: Sodium chloride and sodium chlorate(I) are formed:

$$Cl_2(aq) + 2NaOH(aq) \rightarrow NaCl(aq) + NaClO(aq) + H_2O(l)$$

OxNo: 0 −1 +1

With hot concentrated alkali: Sodium chloride and sodium chlorate(V) are formed:

$$3Cl_2(aq) + 6NaOH(aq) \rightarrow 5NaCl(aq) + NaClO_3(aq) + 3H_2O(l)$$

OxNo: 0 −1 +5

Both these reactions are **disproportionation reactions** (reactions in which a single substance becomes oxidised to a higher oxidation state and reduced to a lower oxidation state). In these examples chlorine has been reduced to Cl^- ions and oxidised to either chlorate(I) or chlorate(V) ions.

Oxidation number has been abbreviated to OxNo in the above equations.

Key points

- Halides react with aqueous silver nitrate to form characteristically coloured precipitates which dissolve to different extents in aqueous ammonia.
- Halide ions are increasingly good reducing agents in the order $Cl^- < Br^- < I^-$ as exemplified by their ability to reduce concentrated sulphuric acid.
- Chlorine reacts with cold and hot aqueous solutions to form chlorate(I) and chlorate(V) ions respectively. These are disproportionation reactions. Chloride ions are formed in both these reactions.

Revision questions

1 Which of the following best describes the order of decreasing solubility of the Group II metal hydroxides?

A Ba > Sr > Ca > Mg > Be

B Ba > Sr > Ca > Be > Mg

C Sr > Ca > Ba > Mg > Be

D Be > Mg > Ca > Sr > Ba

2 Which of the following statements relating to the stability of the +2 and +4 oxidation states of the Group IV elements is correct?

A Tin(II) compounds are readily oxidised to tin(IV).

B Lead(IV) compounds are easily oxidised to lead(II).

C The stability of the +2 oxidation state increases with metallic character.

D Germanium exhibits semi-conductor properties.

3 As the elements in Period 3 of the Periodic Classification is traversed from left to right, which of the following statements is true?

A Electronegativity decreases.

B Melting points rise steadily.

C Atomic radii decrease.

D Screening of valence electrons increases.

4 The relative acid strength of the hydrogen halides, HX, is represented by the following order:

$$HF < HCl < HBr < HI$$

Which of the following statements provides the best explanation for this order?

A The thermal stability of HX increases as the group is descended.

B Hydrogen iodide has the largest standard bond dissociation enthalpy.

C The boiling point of HX increases as the group is ascended.

D Hydrogen bonding decreases as the group is descended.

5 Anhydrous aluminium chloride is a covalently bonded compound.
Which of the following statements correctly explains this statement?

A The radius of the Al^{3+} ion is large.

B The charge density of the Al^{3+} ion distorts the electron charge of the Cl^- ion.

C The electronegativity difference between Al^{3+} and Cl^- ions is very high.

D The Cl^- ion effectively polarises the Al^{3+} ion.

Answer question 6 after reading pages 162–7

6 **a** Explain the terms 'monodentate' and 'bidentate' as applied to ligands, and give an example of each showing, by use of an arrow, the site from which bonding can occur.

b **i** Give the name of a complex ion in which the metal atom exhibits a co-ordination number of 4 and 6 respectively.

ii Sketch the shape of the ions in **i** showing the spatial arrangement of bonds.

c The chloride ion, Cl^-, is able to displace water in the blue hexaaquacopper(II) ion to produce the green tetrachlorocuprate(II) ion.

i Write an equation for this displacement.

ii State the co-ordination number of the copper atom in each ion.

iii Suggest a reason for the difference in co-ordination number in the tetrachloro-complex.

7 **a** Suggest reasons for, and give examples of, the steady rise in the maximum oxidation states across the Period 3 elements.

b The pH of aqueous solutions of the chlorides of the first three elements of Period 3 is given below:

Element	**Na**	**Mg**	**Al**
pH	7	7	5

Use a knowledge of structure and bonding to explain the above information.

c Describe the ease of reaction between the elements sodium, aluminium and chlorine with water, giving relevant equations.

8 a The table below shows the densities of the elements in Group 3, sodium to chlorine:

Element	Na	Mg	Al	Si	P	S	Cl
Density/ $g\,cm^{-3}$	1.0	1.7	2.7	2.3	1.8	2.1	< 0.1

Draw a graph to indicate the variation of density across Period 3, elements sodium to chlorine.

b Refer to their structure and bonding to explain the:

i steady rise of density between sodium and aluminium

ii large decrease in density between sulphur and chlorine.

9 State reasons and give an example, using an equation whenever possible, for the observations in (a)–(c):

a the tetrahalides, MCl_4, of the Group IV elements:

i are volatile, non-polar compounds

ii increase in their reaction with water as the group is descended.

b Carbon monoxide, CO, is a better reducing agent than carbon dioxide, CO_2.

c Silicon(IV) oxide has a very high melting point but does not conduct electricity. Use structure and bonding to give an explanation for this statement about the oxide.

d Copy and complete the following table by inserting the ions formed when the oxides react with **i** acid and **ii** alkali.

Oxide reactions	**i** with acid	**ii** with alkali
Germanium(II) oxide		
Tin(II) oxide		
Lead(IV) oxide		

10 a The table below indicates the decomposition temperatures of the four Group (II) carbonates:

Carbonate	$MgCO_3$	$CaCO_3$	$SrCO_3$	$BaCO_3$
Temperature/K	813	1173	1553	1633

Carefully explain the trend shown.

b i State the trend of solubility of the Group II sulphates.

ii Use information provided by the relevant enthalpy changes to explain this trend.

c List one commercial use for each of the following:

i calcium oxide

ii magnesium oxide

iii calcium carbonate

d Use your knowledge of Group II chemistry to predict the following properties of radium:

i reaction with water, and the subsequent pH of the resulting solution

ii the solubility of the chloride and carbonate

iii the ease of decomposition of the nitrate(v) and its products.

11 a Explain with reference to structure and bonding the variation of melting point in the Group VII elements.

b i State what you would see if bromine were to be added to solutions of potassium iodide and potassium chloride respectively.

ii Name the type of reaction involved.

iii How would you explain your observations in respect to your answer to part **ii**?

c A black powder, P, containing manganese when heated with conc. HCl evolved a greenish gas ,Q, which dissolved in water. A solution of this gas, S, turned litmus paper red before being bleached.

S when allowed to stand in an inverted test tube over water in a beaker produced bubbles of a gas, T, on exposure to sunlight.

i Suggest a name and formula for P.

ii Name the gas Q.

iii Write an equation to explain the production of Q.

iv Write an equation for the reaction producing solution S.

v Deduce the oxidation states of the halogen in the anions found in S.

vii Name the type of reaction which produced S.

viii Write an equation for the production of the gas T.

13 Transition elements

13.1 An introduction to transition elements

Learning outcomes

On completion of this section, you should be able to:

- describe the electronic configuration of the first row transition elements and their ions
- describe the oxidation states of the first row transition elements
- describe the characteristics of transition elements.

Electronic configuration of transition elements

A **transition element** is a d block element which forms one or more stable ions with an incomplete d electron sub-shell. The electron configurations of the first row transition elements are shown below:

Element	Electronic configuration
titanium, Ti	$1s^22s^22p^63s^23p^63d^24s^2$
vanadium, V	$1s^22s^22p^63s^23p^63d^34s^2$
chromium, Cr	$1s^22s^22p^63s^23p^63d^54s^1$
manganese, Mn	$1s^22s^22p^63s^23p^63d^54s^2$
iron, Fe	$1s^22s^22p^63s^23p^63d^64s^2$
cobalt, Co	$1s^22s^22p^63s^23p^63d^74s^2$
nickel, Ni	$1s^22s^22p^63s^23p^63d^84s^2$
copper, Cu	$1s^22s^22p^63s^23p^63d^{10}4s^1$

Did you know?

Scandium, Sc, and zinc, Zn, are not transition elements although they are d block elements. Sc forms only one ion, Sc^{3+}, with no electrons in its d sub-shell. Zn forms only one ion, Zn^{2+}, and has a complete 3d sub-shell.

The electronic configurations of Cr and Cu do not follow the expected pattern of filling the d sub-shell. For Cr, the arrangement of a d sub-shell with one electron in each orbital, $[Ar]3d^54s^1$, gives a greater stability than having one of the d orbitals completely filled. For Cu, the arrangement of a d sub-shell with paired electrons in each orbital, $[Ar]3d^{10}4s^1$, gives a greater stability than having one of the d orbitals with only a single electron in it.

When transition elements lose electrons to form ions, it is the 4s electrons which are lost first, e.g.

Ti atom: $1s^22s^22p^63s^23p^63d^24s^2$ Ti^{2+} ion: $1s^22s^22p^63s^23p^63d^2$
V atom: $1s^22s^22p^63s^23p^63d^34s^2$ V^{3+} ion: $1s^22s^22p^63s^23p^63d^2$

Transition elements can form more than one type of ion, e.g.

Fe^{2+}: $1s^22s^22p^63s^23p^63d^6$ Fe^{3+}: $1s^22s^22p^63s^23p^63d^5$

The range of oxidation states

Transition elements may have several different oxidation states.

- The most common oxidation state of transition elements is +2.
- The maximum oxidation state of the transition elements up to Mn involves all the 4s and 3d electrons.
- From Fe onwards, the +2 oxidation state dominates because the 3d electrons become increasingly harder to remove as the nuclear charge increases.
- Higher oxidation states of transition elements are found in **complex ions** (see Section 13.3) or compound ions such as MnO_4^- and CrO_4^{2-}.

Ti 1 2 ③④
V 1 ②③④⑤
Cr 1 2 ③ 4 5 ⑥
Mn 1 ② 3 ④ 5 ⑥⑦
Fe 1 ②③ 4 5 6
Co 1 ②③ 4 5
Ni 1 ② 3 4
Cu ①② 3

Figure 13.1.1 *The range of oxidation states in transition element compounds in Period 4. All oxidation states are positive. The commonest oxidation states are ringed.*

Characteristics of transition elements

Transition elements have typical metallic properties but have some properties which set them apart from other metals:

- They form compounds with different oxidation states.
- Transition element ions form coloured compounds.
- Transition element ions form complex ions.
- Transition elements and their compounds are often good catalysts (this is often related to the differences in oxidation states).
- Transition elements have very high density.
- Transition elements have very high melting points and boiling points.
- They have typical magnetic properties:

Most compounds and ions of the transition elements are **paramagnetic**: when placed in a magnetic field, they align themselves with the field. But they do not retain their magnetism when the magnetic field is removed.

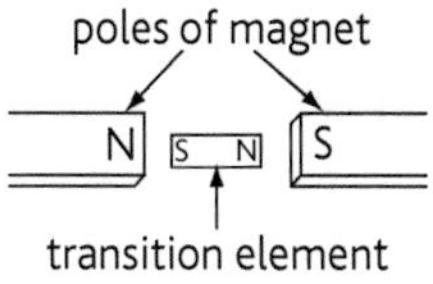

Figure 13.1.2 *A small rod of a transition element aligns itself in a magnetic field*

Fe, Co and Ni are **ferromagnetic**. They retain permanent magnetism after the magnetic field has been withdrawn.

Key points

- Transition elements form one or more ions with an incomplete d sub-shell of electrons.
- When a transition element forms an ion, the electrons are first lost from the 4s sub-shell and then from the 3d sub-shell.
- Transition elements exist in various oxidation states, form coloured compounds (containing complex ions), and often have characteristic magnetic properties and catalytic activity.

13.2 Physical properties of transition elements

Learning outcomes

On completion of this section, you should be able to:

- explain the relatively small changes in atomic and ionic radius and first ionisation energy of the transition elements across a period
- describe qualitatively the physical properties of transition elements compared with calcium.

Ionisation energy, atomic and ionic radii

In Periods 2 and 3 of the Periodic Table there is a considerable variation in the values of the first ionisation energy and the atomic and ionic radii. The transition elements, however, show only small changes in these values from titanium to copper.

Element	Ti	V	Cr	Mn	Fe	Co	Ni	Cu
$\Delta H^{\ominus}_{i1}$/kJ mol^{-1}	661	648	653	716	762	757	736	745
Atomic radius/nm	0.132	0.122	0.117	0.117	0.116	0.116	0.115	0.117
Ionic radius of 2+ ions/nm	0.090	0.090	0.085	0.080	0.076	0.078	0.078	0.069

Across this series of transition elements, the first ionisation energy, $\Delta H^{\ominus}_{i1}$, generally increases, but only by a very small amount. Atoms of each successive element have one more proton in their nucleus. This should increase the attraction of the outer electrons markedly since the added electrons are going into the same sub-shell. But there is an increased repulsion on the outer electrons caused by the additional electron going into the d sub-shell. The effect of adding electrons to a d sub-shell, in the case of transition elements, is to make the difference in ionisation energy between one element and the next much smaller. Similarly, the atomic and ionic radii tend to increase along the series, but only slightly compared with the large differences in the s and p block elements. The increased attractive effect of the nucleus on the outer electrons from one transition element to the next is almost cancelled by the increased repulsive forces caused by the extra repulsion of the 3d electrons.

Comparing transition elements with calcium

Calcium is the s block element in Group II which is placed just before the d block elements in Period 4. Although calcium and the transition elements of Period 4 are metals, there are some marked differences between them. The table compares some properties of calcium with two transition elements, iron and nickel.

	calcium	iron	nickel
Melting point/°C	839	1540	1450
Density/g cm^{-3}	1.55	7.86	8.90
Atomic radius/nm	0.197	0.116	0.115
Ionic radius/nm	0.099	0.076	0.078
First ionsation energy/kJ mol^{-1}	590	762	736

The transition elements are harder and have higher melting points than calcium. This reflects the stronger metallic bonding in transition elements compared with calcium. Calcium can only release its two outer s electrons to form the delocalised 'sea of electrons' but transition elements can release electrons from both the 4s and 3d sub-levels. There is a greater force of attraction between the small (and often highly charged) transition element ions and the sea of electrons in comparison with calcium.

The atomic and ionic radii of calcium are much larger than the corresponding radii for typical transition elements. This is related to the presence of the filling of the d sub-shell in the transition elements. Electrons in the d sub-shell are less good at shielding the outer electrons from the nuclear charge than s or p electrons. So there is a relatively greater force of attraction between the nucleus and the outer electrons which tends to pull the electrons closer to the nucleus.

The first ionisation energies of transition elements are higher than that of calcium because of this relatively greater force of attraction between the nucleus and the outer electrons.

The transition elements are denser than calcium. This is due to the fact that transition element ions have smaller radii than calcium.

The electronegativity of all the transition elements is significantly higher compared with calcium, e.g. Ca = 1.0, Fe and Ni = 1.8. The electronegativity increases across the series from Ti to Cu as the elements get slightly less metallic in character.

Electrical conductivity of transition elements

The table compares the electrical conductivity of calcium with some of the transition elements.

Element	Ca	Ti	Mn	Fe	Ni	Cu
Conductivity/10^8 S m^{-1}	0.30	0.024	0.007	0.10	0.15	0.63

Most of the transition elements are good conductors of electricity. Those with a single outer s electron and a half filled or completely full d shell are especially good conductors. So copper and chromium are relatively better conductors than manganese and nickel. Although calcium is a good electrical conductor, it is too reactive to be used in electrical wiring.

Key points

- Across the first row of transition elements, the changes in atomic and ionic radius and first ionisation energy are relatively small. This is because each electron added goes into a d sub-shell and so the shielding effect is minimised.
- The physical properties of transition elements differ from those of a typical s block element because of the presence of the d sub-shell.

13.3 Coloured ions and oxidation states

Learning outcomes

On completion of this section, you should be able to:

- describe the various oxidation states of vanadium
- explain the formation of coloured ions by transition elements.

The various oxidation states of vanadium

Transition elements can form ions in more than one oxidation state. Vanadium is one such element, whose ions in their various oxidation states exhibit characteristic colours. Vanadium (V) is usually supplied as solid ammonium vanadate, NH_4VO_3. When ammonium vanadate is acidified the vanadium becomes part of a positive ion, VO_2^+, in which vanadium has an oxidation state of +5.

oxidation state ↑		
+5	VO_2^+	yellow
+4	VO^{2+}	blue
+3	V^{3+}	green
+2	V^{2+}	purple/mauve
0	V	

The redox chemistry of vanadium can be explained using an electrode potential chart.

$E^{\ominus}$/V	
−0.76	$Zn^{2+}_{(aq)} + 2e^- \rightleftharpoons Zn_{(s)}$
−0.26	$V^{3+}_{(aq)} + e^- \rightleftharpoons V^{2+}_{(aq)}$
−0.14	$Sn^{2+}_{(aq)} + 2e^- \rightleftharpoons Sn_{(s)}$
+0.34	$VO^{2+}_{(aq)} + 2H^+_{(aq)} + e^- \rightleftharpoons V^{3+}_{(aq)} + H_2O_{(l)}$
+0.77	$Fe^{3+}_{(aq)} + e^-_{(aq)} + e^- \rightleftharpoons Fe^{2+}_{(aq)}$
+1.00	$VO_2{}^+_{(aq)} + 2H^+_{(aq)} + e^- \rightleftharpoons VO^{2+}_{(aq)} + H_2O_{(l)}$

Figure 13.3.1 *An electrode potential chart showing the oxidation states of vanadium and other compounds*

By reference to $E^{\ominus}$ values, we can select suitable compounds to reduce vanadium(v) ions to a lower oxidation state:

- Addition of Fe^{2+} ions to VO_2^+ ions will reduce the VO_2^+ ions to VO^{2+} ions. Fe^{2+} ions are better reducing agents than VO^{2+} ions and VO_2^+ ions are better oxidants than Fe^{3+} ions. So the reaction $VO_2^+ \rightarrow VO^{2+}$ goes in the forward direction and the reaction with the higher $E^{\ominus}$ value goes in the reverse direction ($Fe^{2+} \rightarrow Fe^{3+}$). The colour change will be from yellow to blue.
- Addition of Sn to VO_2^+ ions will reduce the VO_2^+ ions to V^{3+} ions. Using the anticlockwise rule, the reaction $VO_2^+ \rightarrow VO^{2+} \rightarrow V^{3+}$ goes in the forward direction and the reaction with the higher $E^{\ominus}$ value goes in the reverse direction ($Sn \rightarrow Sn^{2+}$). The colour change will be from yellow to blue to green.

  $$Sn(s) + 2VO_2^+(aq) + 4H^+(aq) \rightarrow 2VO^{2+}(aq) + Sn^{2+}(aq) + 2H_2O(l)$$

  then: $$Sn(s) + 2VO^{2+}(aq) + 4H^+(aq) \rightarrow 2V^{3+}(aq) + Sn^{2+}(aq) + 2H_2O(l)$$

- Addition of Zn to VO_2^+ ions will reduce the VO_2^+ ions to V^{2+} ions. Using the anticlockwise rule, the reaction $VO_2^+ \rightarrow VO^{2+} \rightarrow V^{3+} \rightarrow V^{2+}$ goes in the forward direction and the reaction with the higher $E^{\ominus}$ value goes in the reverse direction ($Zn \rightarrow Zn^{2+}$). The colour change will be from yellow to blue to green to purple.

The formation of coloured ions

Compounds appear coloured when they absorb energy that corresponds to certain wavelengths of light in the visible spectrum. Aqueous Ti^{3+} ions appear purple because they absorb light mostly in the green region of the spectrum. Most of the light which passes through is from the violet, blue and red regions. So the solution appears purple.

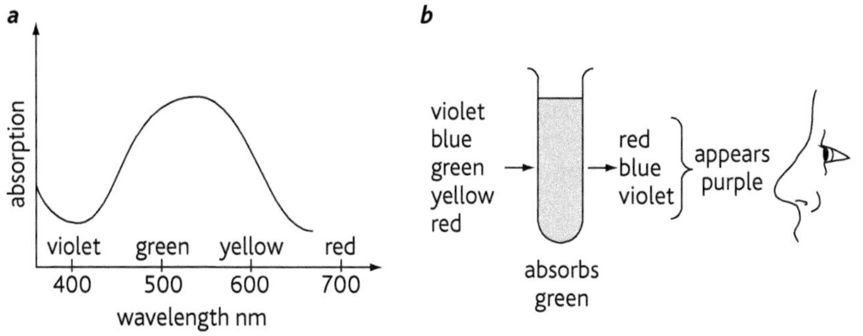

Figure 13.3.2 ***a*** *The absorption spectrum of aqueous Ti^{3+} ions;* ***b*** *Only violet, blue and red light are transmitted through the solution.*

What causes the colour?

Transition element ions, such as Ti^{3+}, in solution are bonded to a definite number of water molecules. Each water molecule forms a co-ordinate (dative covalent) bond with the transition element ion. These water molecules are called **ligands** and the resulting ion is called a **complex ion**. The complex ion in the Ti^{3+}–water complex has the formula $[Ti(H_2O)_6]^{3+}$.

- The 3 d orbitals in an isolated transition element ion are described as **degenerate orbitals**. They all have the same average energy.
- The presence of the ligands in a complex affects the electrons in the d orbitals of the transition element ion. Orbitals close to the ligands are pushed to slightly higher energy levels than those further away. The orbitals split into two groups.
- When an electron moves from a d orbital of lower energy to a d orbital of higher energy, light is absorbed in the visible region of the spectrum.
- The frequency of the light absorbed depends on the energy difference between the split d levels. Different ligands split the d energy levels by different amounts. So different ligands may cause different colours to be absorbed.

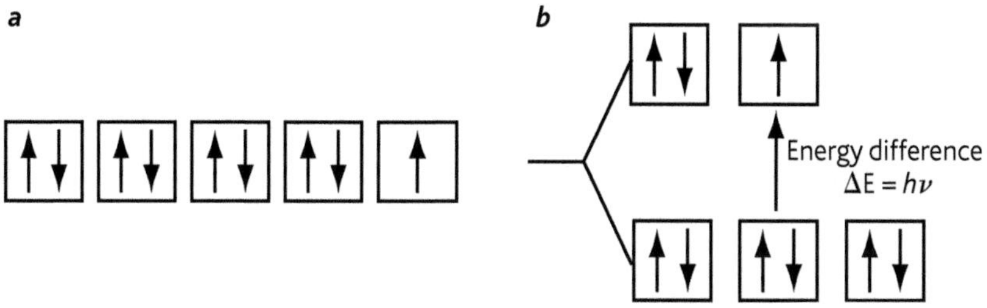

Figure 13.3.3 ***a*** *Degenerate orbitals in a Cu^{2+} ion;* ***b*** *The ligand (water) in a $[Cu(H_2O)_6]^{2+}$ ions causes d orbital splitting*

Key points

- An acidified solution of ammonium vanadate(v) can react with reducing agents to form vanadium ions with different oxidation states depending on the relative $E^{\ominus}$ values for the half reactions.
- A ligand is a molecule or ion with one or more lone pair of electrons which can form a co-ordinate bond with a transition metal ion.
- Transition element compounds are often coloured because of **d-orbital splitting** caused by ligands.

13.4 Complex ions and ligand exchange

Learning outcomes

On completion of this section, you should be able to:

- explain the principle of ligand exchange
- know that ligand exchange depends on the stability constants in an equilibrium reaction where the ligands compete for bonding with the transition element ion
- describe some examples of ligand exchange.

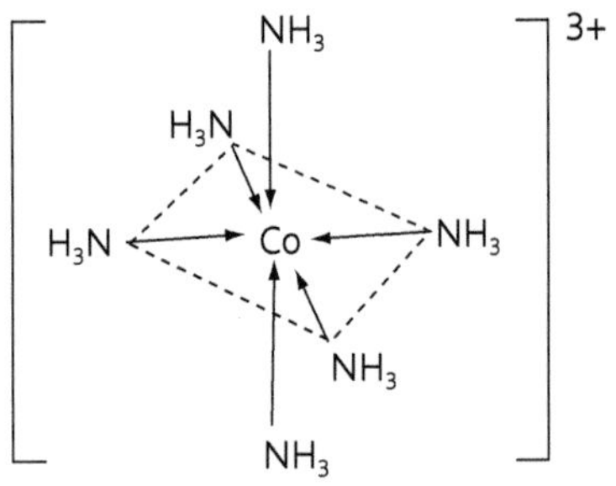

$[Co(NH_3)_6]^{3+}$

Figure 13.4.1 *A complex ion formed from a cobalt 3+ ion and six ammonia molecules*

Exam tips

Remember that the charge on a complex ion is the sum of the charges on the transition metal ion and all the ligands.

More about ligands and complex ions

A **ligand** is a molecule or ion with one or more lone pairs of electrons available to donate to a transition element ion. Examples of simple ligands are water, ammonia and chloride ions. Co-ordinate (dative covalent) bonds are formed with the transition element ions because the d electrons in transition elements do not shield the outer electrons very well from the nuclear charge. So, the relatively poorly shielded nucleus attracts lone pairs of electrons strongly enough to form co-ordinate bonds and a complex ion is formed. Some examples are shown in the table below.

Transition element ion	Ligand	Formula of complex
Fe^{2+}	water, H_2O	$[Fe(H_2O)_6]^{2+}$
Co^{3+}	ammonia, NH_3	$[Co(NH_3)_6]^{3+}$
Cu^{2+}	chloride ion, Cl^-	$[CuCl_4]^{2-}$
Cr^{3+}	hydroxide ion, OH^-	$[Cr(OH)_6]^{3-}$
Ag^+	ammonia, NH_3	$[Ag(NH_3)_2]^+$
Ni^{2+}	diaminoethane, $NH_2CH_2CH_2NH_2$	$[Ni(NH_2CH_2CH_2NH_2)_3]^{2+}$.

Co-ordination number

The **co-ordination number** of a complex ion is the number of co-ordinate bonds a ligand forms with the central transition metal ion.

- **Monodentate** ligands form one bond per ligand, e.g. water, ammonia.
- **Bidentate** ligands form two bonds per ligand, e.g. diaminoethane. These ligands have two lone pairs available to form co-ordinate bonds.
- Hexadentate ligands form 6 bonds per ligand, e.g. EDTA.

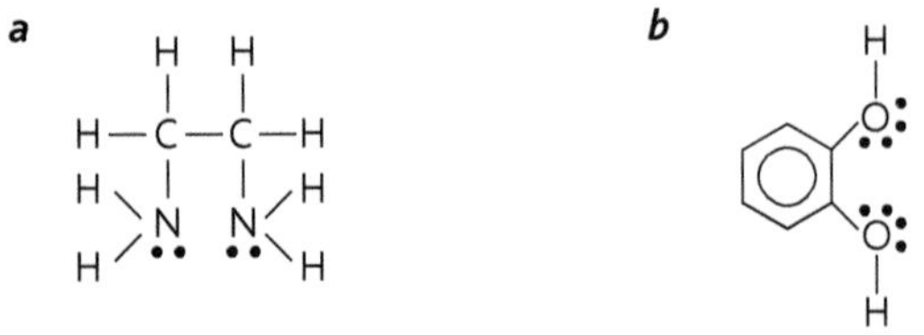

Figure 13.4.2 *Two bidentate ligands:* ***a*** *1,2-diaminoethane;* ***b*** *benzene-1,2-diol*

Ligand exchange

If there is more than one ligand in a solution, they can compete for a transition metal cation. For example, aqueous ammonia contains two ligands: H_2O and NH_3. The better the ligand is at competing for the transition element ion, the more stable is the complex formed. When we add a few drops of concentrated HCl to an aqueous solution of Cu^{2+} ions (which contains the complex ions $[Cu(H_2O)_6V^{2+}]$, the following equilibrium is set up:

$$\underset{\text{blue}}{[Cu(H_2O)_6]^{2+}(aq)} + 4Cl^-(aq) \rightleftharpoons \underset{\text{yellow-green}}{[CuCl_4]^{2-}(aq)} + 6H_2O(l)$$

Addition of HCl to the $[Cu(H_2O)_6]^{2+}$ complex shifts the position of equilibrium to the right and so the colour changes from blue to green as the complex $[CuCl_4]^{2-}(aq)$ is formed. Addition of water shifts the position of equilibrium to the left.

The equilibrium constant for this reaction is called the **stability constant**, K_{stab}.

$$K_{stab} = \frac{[[CuCl_4]^{2-}(aq)]}{[[Cu(H_2O)_6]^{2+}] \times [Cl^-(aq)]^{4-}}$$

The larger the value of the stability constant, the more stable is the complex and the more likely the complex will form.

Ammonia has a higher stability constant than Cl^- ions. So addition of ammonia will shift the position of equilibrium to the right and a deep blue complex ion is formed:

$$\underset{\text{yellow-green}}{[CuCl_4]^{2-}(aq)} + 4NH_3 + 2H_2O(l) \rightleftharpoons \underset{\text{deep blue}}{[Cu(NH_3)_4(H_2O)_2]^{2+}(aq)} + 4Cl^-(aq)$$

Cobalt(II) ions occur as complexes with water, Cl^- ions and ammonia. The ligand exchanges are similar to those of the copper complexes:

$$\underset{\text{pink}}{[Co(H_2O)_6]^{2+}(aq)} + 4Cl^-(aq) \rightleftharpoons \underset{\text{blue}}{[CoCl_4]^{2-}(aq)} + 6H_2O(l)$$

$$\underset{\text{pink}}{[Co(H_2O)_6]^{2+}(aq)} + 6NH_3(aq) \rightleftharpoons \underset{\text{yellow}}{[Co(NH_3)_6]^{2+}(aq)} + 6H_2O(l)$$

Ligand exchange in haem

The red blood pigment haem is found as a group attached to the protein haemoglobin in red blood cells. The molecule has an Fe^{2+} ion with a co-ordination number of 6. Four of the co-ordination positions are to nitrogen atoms in a complex ring system. A fifth co-ordination position is with a nitrogen atom in the protein molecule. The sixth coordination position is with an oxygen molecule. The oxygen molecule is weakly bound and carried to the cells for respiration. Carbon monoxide will also bind to the Fe^{2+} ion. It has a stability constant about 200 times higher than O_2. In the presence of carbon monoxide, hardly any oxygen will bind and respiration is inhibited. The result is often death, even if the concentration of carbon monoxide is fairly low.

Key points

- Transition elements form complexes by combining with one or more ligands.
- Ligands are bonded to the transition element ion by one or more co-ordinate bonds.
- Ligand exchange depends on the stability constants in an equilibrium reaction where the ligands compete for bonding with the transition element ion.
- A colour change is often observed when one or more ligands are exchanged for other ligands in a complex.
- Examples of ligand exchange include $Cu^{2+}(aq)$ or $Co^{2+}(aq)$ with NH_3 and CO/O_2 with the Fe^{2+} complex in haem.

13.5 Complex ions and redox reactions

Learning outcomes

On completion of this section, you should be able to:

- deduce the shapes of complexes (square planar, tetrahedral or octahedral)
- describe the use of Fe^{3+}/Fe^{2+}, MnO_4^-/Mn^{2+} and $Cr_2O_7^{2-}/Cr^{3+}$ as redox systems.

The shapes of complex ions

The shape of a complex ion depends on:

- its co-ordination number
- the type of ligand which bonds to the transition element ion.

The shape of the complex cannot be predicted using the VSEPR theory because the electrons in d orbitals differ from those in s and p orbitals in their influence on structure.

The most common co-ordination numbers are 2, 4 and 6.

Co-ordination number 2

Complexes involving gold, silver or copper in oxidation state +1 usually have a linear structure, e.g. $[Ag(NH_3)_2]^+$, $CuCl_2^-$

$$[H_3N \rightarrow Ag \leftarrow NH_3]^+$$

Co-ordination number 4

These may be either square planar or tetrahedral. The square planar shape is more common. It is found in many nickel(II), copper(II) and platinum(II) complexes. The tetrahedral form is found in the $[CoCl_4]^{2-}$ ion and the $Ni(CO)_4$ molecule. The reason why the $[CoCl_4]^{2-}$ ion has a coordination number of 4 rather than 6 (see Figure 13.4.1) is that the chloride ion is relatively large. So fewer ligands can fit round the central transition element ion.

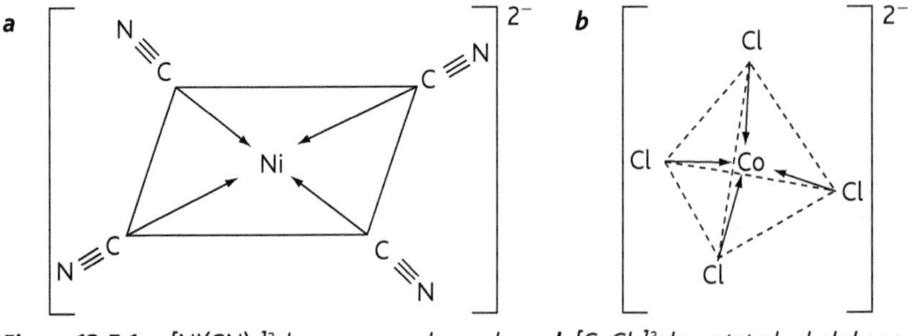

Figure 13.5.1 ***a*** *$[Ni(CN)_4]^{2-}$ has a square planar shape;* ***b*** *$[CoCl_4]^{2-}$ has a tetrahedral shape*

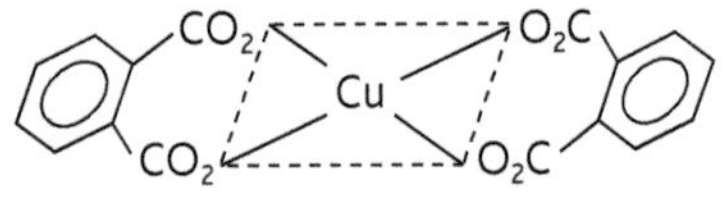

Figure 13.5.2 A square planar complex of copper with 2-hydroxybenzoate ions

Square planar complexes can also be formed using bidentate ligands. Each ligand forms two coordinate bonds with the transition element ion.

Co-ordination number 6

This is the commonest type of coordination. It is always octahedral. Examples are:

$$[Co(H_2O)_6]^{3+},\ [Fe(CN)_6]^{3-} \text{ and } [Ni(NH_3)_6]^{2+}.$$

An octahedral shape is usually formed if the ligand atom forming the bond with the central transition element ion is relatively small. Octahedral complexes can also be formed using three bidentate ligands. Each ligand forms two coordinate bonds with the transition element ion.

Did you know?

Doctors sometimes use the hexadentate ligand EDTA as an antidote to transition metal poisoning. The complex ions are excreted in the urine.

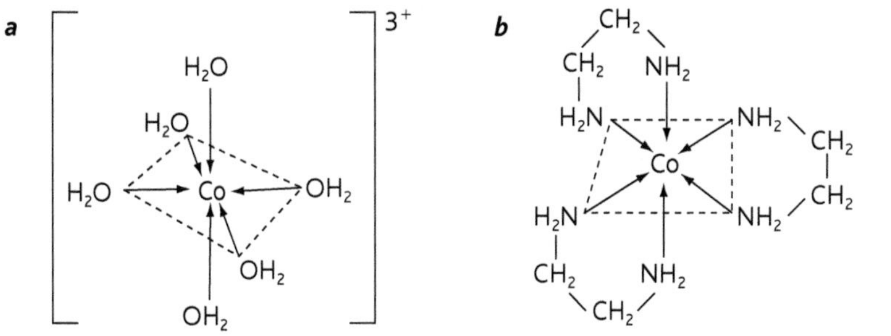

***Figure 13.5.3** Octahedral complexes of cobalt: **a** with a monodentate ligand, water; **b** with a bidentate ligand, diaminoethane*

Transition element compounds and redox

When compounds of transition elements are treated with suitable reagents, the oxidation state of the transition element may change. Different oxidation states of transition element complexes are often useful in demonstrating redox reactions. A solution containing iron(II) ions such as a solution of iron(II) ammonium sulphate can be used in such a way. A few drops of oxidising agent such as acidified potassium manganate(VII) will oxidise iron(II) ions to iron(III) ions.

$$5Fe^{2+}(aq) + MnO_4^-(aq) + 8H^+(aq) \rightarrow 5Fe^{3+}(aq) + Mn^{2+}(aq) + 4H_2O(l)$$

The colour of the solution changes from the light green of the Fe^{2+} ions to the reddish-brown colour of Fe^{3+} ions. The addition of alkali to the products confirms the presence of iron(III) ions when a red-brown precipitate is formed. The opposite colour change occurs when reducing agents such as zinc react with iron(III) ions to form iron(II) ions e.g.

$$2Fe^{3+}(aq) + Zn(s) \rightarrow 2Fe^{2+}(aq) + Zn^{2+}(aq)$$

The iron(II) ions formed in this reaction can be tested for by adding alkali to the solution when a grey-green precipitate is formed.

Potassium managnate(VII)

The MnO_4^- ions present in a solution of acidified potassium manganate(VII) can be used as an oxidising agent to test for sulphur dioxide and hydrogen sulphide as well as being useful in redox titrations (see page 45). It is reduced to Mn^{2+} ions. The colour change can be used to indicate the end point of a redox titration. At the end point a permanent pink colouration is seen due to the presence of excess MnO_4^- ions. An example is in estimating the concentration of ethanedioic (oxalic acid), $H_2C_2O_4$.

$$5C_2O_4^{2-}(aq) + 2MnO_4^-(aq) + 16H^+(aq) \rightarrow 10CO_2(g) + 2Mn^{2+}(aq) + 8H_2O(l)$$

Potassium dichromate(VI)

Potasium dichromate(VI) is a good oxidizing agent. In acidic conditions, the orange $Cr_2O_7^{2-}$ ions are reduced to green Cr^{3+} ions. It is used as in volumetric analysis to determine the concentration of Fe^{2+} ions.

$$Cr_2O_7^{2-}(aq) + 6Fe^{2+}(aq) + 14H^+(aq) \rightarrow 2Cr^{3+}(aq) + 6Fe^{3+}(aq) + 7H_2O(l)$$

A redox indicator such as barium diphenylamine sulphonate is added. The first drop of excess dichromate converts this to a deep bluish solution.

Did you know?

Potassium dichromate is usually preferred to potassium manganate(VII) in redox titrations because it can be obtained to a high degree of purity. It is used as a primary standard because it is also stable in air, readily soluble to give a stable solution, has a high molar mass and titrates reproducibly.

Key points

- The shape of transition element complexes depends on the number and type of ligand which bonds to the metal cation.
- The shapes of complexes are linear, tetrahedral, square planar and octahedral.
- MnO_4^- and $Cr_2O_7^{2-}$ can be used as oxidising agents and to test for reducing agents in the laboratory. During these reactions characteristic colour changes occur.

14 Qualitative tests for ions

14.1 Identifying cations (1)

Learning outcomes

On completion of this section, you should be able to:

- describe how a flame test can be used to identify particular metal cations
- describe how to identify specific cations using an aqueous solution of carbonate ions, know the principles on which this is based and write equations for these reactions
- describe a test for ammonium ions.

Qualitative analysis

There are four stages in analysing the elements present in a compound:

- Preliminary tests: these include the appearance of the solid, flame test and the action of heat on the compound.
- Making a solution of the substance under test.
- Testing for the cations present.
- Testing for the anions present. (See Section 14.3)

The flame test

A flame test can be used to identify some cations, especially those in the s block of the Periodic Table. The procedure is:

- Clean a platinum or Nichrome wire by dipping it in concentrated hydrochloric acid.
- Place a sample of the compound on the end of the wire.
- Hold the wire on the edge of a non-luminous Bunsen flame.
- Note any colour in the flame.
- Observe the coloured flame through a diffraction grating or spectroscope.

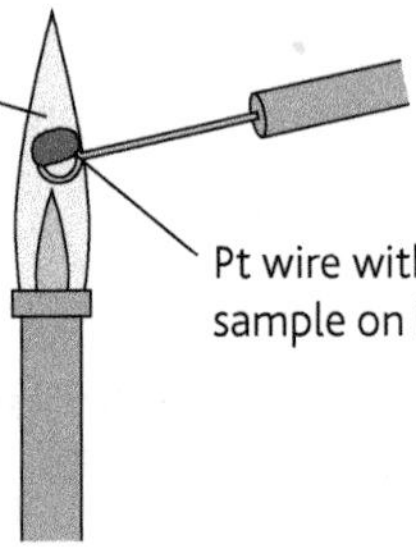

Figure 14.1.1 *Carrying out a flame test*

The typical flame colours for selected metal cations are:

Group I	Group II	Transition element
sodium – orange-yellow	calcium – brick red (orange-red)	copper – greenish-blue
potassium – lilac	strontium – crimson red	
	barium – apple-green	

When you look at the flame through a diffraction grating or spectroscope, you will see the coloured lines of the line emission spectrum in the visible region (see Section 1.3). The colours seen in the flame test are due to the most obvious lines in the line emission spectrum. For example, the yellow lines in the sodium spectrum are particularly distinct.

Making a solution of the solid

- Attempt to dissolve the solid in a few cubic centimetres of water.
- If it does not dissolve, warm gently.
- If it is insoluble in water, try to dissolve it in nitric acid.

Many tests for cations and anions depend on solubility and precipitation. So it is important to know the solubility rules for common inorganic compounds:

- Common salts of Group I cations and NH_4^+ are soluble.
- All nitrates are soluble.
- Sulphates are soluble except those of Ca^{2+}, Ba^{2+}, Sr^{2+} and Pb^{2+}.
- Chlorides, bromides and iodides are soluble except those of Ag^+ and Pb^{2+}.
- Carbonates and sulphites are insoluble except those of Group I cations and NH_4^+.
- All hydroxides are insoluble except those of Group I cations, NH_4^+, Sr^{2+} and Ba^{2+}.

Using sodium carbonate to test for cations

Aqueous sodium carbonate is sometimes used to confirm the presence or absence of a Group I cation or NH_4^+. Most carbonates are insoluble in water.

- An aqueous solution of Group I cations or NH_4^+ will not form a precipitate on addition of sodium carbonate because these carbonates are soluble.
- An aqueous solution containing other metal cations will give a precipitate because their carbonates are insoluble. Group II carbonates give a white precipitate and transition element carbonates may give coloured precipitates, e.g. $MnCO_3$ is pink.
- Mg^{2+} ions can be distinguished from other Group II ions by adding ammonium carbonate in the presence of ammonium chloride. Under these conditions, no precipitate is formed.

A test for ammonium ions

Many ammonium salts decompose on warming. Ammonia gas is given off, e.g.

$$NH_4Cl(s) \rightarrow NH_3(g) + HCl(g)$$
$$(NH_4)_2SO_4(s) \rightarrow 2NH_3(g) + SO_3(g) + H_2O(l)$$

Ammonium nitrate is an exception. It decomposes to form nitrogen(I) oxide rather than ammonia.

All ammonium salts give off ammonia when warmed with dilute aqueous sodium hydroxide:

$$NH_4Cl(aq) + NaOH(aq) \rightarrow NH_3(g) + NaCl(aq) + H_2O(l)$$

The ammonia can be detected by the fact that it turns damp red litmus paper blue.

Did you know?

When ammonium salts are heated in a test tube they almost always decompose. But when the vapours are cooled further up the tube, the solid salt is reformed, e.g.

$$NH_3 + HCl \rightarrow NH_4Cl(s)$$

These compounds can therefore be purified by sublimation.

Key points

- When some elements are put in a non-luminous Bunsen flame, characteristic colours are seen, depending on the main colours in the line emission spectrum.
- Specific metal cations can be identified using an aqueous solution of carbonate ions.
- When a compound containing ammonium ions is warmed with aqueous sodium hydroxide, ammonia is released.

14.2 Identifying cations (2)

Learning outcomes

On completion of this section, you should be able to:

- describe how to identify metal cations using sodium hydroxide or aqueous ammonia
- explain the principles on which the above test is based
- deduce ionic equations with state symbols for the reaction of metal cations with (i) sodium hydroxide (ii) aqueous ammonia.

Exam tips

A lot of memory work is required to learn all these qualitative tests. One way of doing this is to write the ion and test on a sheet of paper in separate columns, then cover one of the columns up and see how well you can remember the tests.

Did you know?

The reaction of some transition metal cations with sodium hydroxide is quite complicated because of complex formation. The actual reaction of Cu^{2+} ions with OH^- ions may be more accurately represented as:

$$[Cu(H_2O)_6]^{2+} + 2OH^- \rightarrow [Cu(H_2O)_4(OH)_2]^{2+} + 2H_2O(l)$$

Exam tips

The formulae of aluminates, zincates and plumbates are sometimes written AlO_2^-, ZnO_2^{2-} and PbO_2^{2-} respectively. You should be prepared to use these formulae as well.

Tests using aqueous sodium hydroxide

Many metal cations can be identified by observing the colour of the precipitate (if any) formed by the addition of dilute sodium hydroxide to an aqueous solution of the substance under test. Some of the precipitates may redissolve in excess sodium hydroxide.

- If no precipitate is formed: A cation from Group I or ammonium ions may be present. (The Group I cations can be distinguished by flame tests and NH_4^+ ions by warming the alkaline solution.)
- If a white precipitate forms which is insoluble in excess sodium hydroxide:

  Mg^{2+}, Ca^{2+}, Sr^{2+} or Ba^{2+} may be present (these can be distinguished by a flame test).
- If a white precipitate forms which is soluble in excess sodium hydroxide:

  Al^{3+}, Pb^{2+} or Zn^{2+} ions may be present.
- If a coloured precipitate is formed the colour may be used to identify the cation:

  Cr^{3+} → grey-green precipitate
  Cu^{2+} → pale blue precipitate
  Fe^{2+} → green precipitate (turning brown)
  Fe^{3+} → rusty-brown (reddish-brown) precipitate
  Mn^{2+} → cream precipitate (turning dark brown)
  Ni^{2+} → green precipitate

The precipitates formed are of metal hydroxides. Metal hydroxides are insoluble in water apart from those of Group I or ammonium ions. Typical equations are:

$$Fe^{2+}(aq) + 2OH^-(aq) \rightarrow Fe(OH)_2(s)$$
$$Cu^{2+}(aq) + 2OH^-(aq) \rightarrow Cu(OH)_2(s)$$
$$Cr^{3+}(aq) + 3OH^-(aq) \rightarrow Cr(OH)_3(s)$$

The precipitates of aluminium, zinc and lead hydroxides dissolve in excess sodium hydroxide because soluble aluminates, zincates and plumbates are formed:

$$Al(OH)_3(s) + NaOH(aq) \rightarrow \underset{\text{sodium aluminate}}{NaAl(OH)_4(aq)}$$

$$Zn(OH)_2(s) + 2NaOH(aq) \rightarrow \underset{\text{sodium zincate}}{Na_2Zn(OH)_4(aq)}$$

$$Pb(OH)_2(s) + 2NaOH(aq) \rightarrow \underset{\text{sodium plumbate(II)}}{Na_2Pb(OH)_4(aq)}$$

Tests using aqueous ammonia

The effect of aqueous ammonia on metal cations in solution provides confirmatory tests for some ions. The colour of the precipitate (if any) formed by the addition of aqueous ammonia to an aqueous solution of the substance under test and the effect of excess aqueous ammonia on the precipitate allows some distinction between cations to be made. Most of the reactions are the same as for sodium hydroxide because aqueous ammonia contains OH^- ions due to the reaction:

$$NH_3(aq) + H_2O(l) \rightleftharpoons NH_4^+(aq) + OH^-$$

- The concentration of hydroxide ions in dilute aqueous ammonia is low in comparison with NaOH, so test solutions containing Ca_2^+ ions, Sr_2^+ and Ba_2^+ ions may form only a slight white precipitate or even no precipitate.
- Al^{3+} can be distinguished from Zn^{2+} by the use of aqueous ammonia. Al^{3+} ions form a white precipitate which is insoluble in excess aqueous ammonia. Zn^{2+} ions form a white precipitate which redissolves in excess aqueous ammonia. A colourless complex ion which is soluble in water is formed:

  $$Zn^{2+}(aq) + 2OH^-(aq) \rightarrow Zn(OH)_2(s)$$

  in excess: $Zn(OH)_2(s) + 4NH_3(aq) \rightarrow [Zn(NH_3)_4]^{2+}(aq) + 2OH^-(aq)$
- Copper hydroxide dissolves in excess ammonia to form complex ions which are soluble in water. The complex ion is a deep blue colour.

  $$Cu^{2+}(aq) + 2OH^-(aq) \rightarrow Cu(OH)_2(s)$$

  in excess:

  $$Cu(OH)_2(s) + 4NH_3(aq) + 2H_2O(l) \rightarrow [Cu(NH_3)_4(H_2O)_2]^{2+}(aq) + 2OH^-(aq)$$

  often simplified to:

  $$Cu(OH)_2(s) + 4NH_3(aq) \rightarrow [Cu(NH_3)_4]^{2+}(aq) + 2OH^-(aq)$$
- Other transition element hydroxides such as $Co(OH)_2$ and $Ni(OH)_2$ will also redissolve in excess aqueous ammonia to form complexes, but $Mn(OH)_2$ does not redissolve.

Key points

- Some metal cations in aqueous solution form characteristically coloured precipitates on addition of NaOH(aq) or NH_3(aq).
- Depending on the metal cation, these precipitates may or may not dissolve in excess NaOH(aq) or NH_3(aq).
- Aluminium and zinc hydroxide redissolve in excess sodium hydroxide and copper(II) hydroxide redissolves in excess ammonia because soluble complex ions are formed.
- Ionic equations can be written to show precipitation and complex ion formation.

14.3 Identifying anions

Learning outcomes

On completion of this section, you should be able to:

- describe tests to identify CO_3^{2-}, NO_3^-, SO_3^{2-}, SO_4^{2-}, Cl^-, Br^-, I^- and CrO_4^{2-} ions
- explain the principles on which these tests are based
- deduce ionic equations with state symbols for these reactions.

Exam tip

You will not be expected to remember complex equations such as the ones for the nitrate test. But you should (i) know the formulae for simple compounds (ii) be able to work out equations from first principles given suitable information (see Section 4.2).

Testing for carbonates

When an acid is added to a carbonate, carbon dioxide is released:

$$CaCO_3(s) + 2HCl(aq) \rightarrow CaCl_2(aq) + CO_2(g) + H_2O(l)$$

Carbon dioxide turns limewater (a dilute solution of calcium hydroxide) milky. The cloudiness in the solution is a fine precipitate of calcium carbonate:

$$Ca(OH)_2(aq) + CO_2(g) \rightarrow CaCO_3(s) + H_2O(l)$$

Testing for nitrates

There are several tests for nitrates:

- Warm a little of the solid with a 2 cm³ of concentrated sulphuric acid. Then add a small piece of copper. If brown fumes of nitrogen dioxide, NO_2, are formed, a nitrate is probably present:

  $$Cu(s) + 2NO_3^-(s) + 4H^+(aq) \rightarrow Cu^{2+}(aq) + 2NO_2(g) + 2H_2O(l)$$

- A confirmatory test is to add aqueous sodium hydroxide to the suspected nitrate and then either zinc powder or aluminium powder (or Devarda's alloy). On warming, ammonia gas is released:

  $$NO_3^-(aq) + 4Zn(s) + 7OH^-(aq) + 6H_2O(l) \rightarrow NH_3(g) + 4Zn(OH)_4^{2-}(aq)$$

  The ammonia can be identified by using damp red litmus paper. If ammonia is present, the litmus turns blue.

Testing for sulphates

The solution to be tested is acidified with nitric acid to remove any contaminating carbonates present. Aqueous barium chloride or aqueous barium nitrate are then added. If a sulphate is present, a white precipitate of barium sulphate is formed (see Section 12.2).

$$Ba^{2+}(aq) + SO_4^{2-}(aq) \rightarrow BaSO_4(s)$$

Testing for sulphites

The sulphite ion has the formula SO_3^{2-}. Sulphites are more unstable to heat than are sulphates so a simple test is to heat a sample of the solid sulphite. Sulphur dioxide gas is released.

If the sulphite is in solution, hydrochloric acid is first added to the solution and the solution is then heated. The release of sulphur dioxide gas indicates the presence of a sulphite:

$$SO_3^{2-}(aq) + 2H^+(aq) \rightarrow SO_2(g) + H_2O(l)$$

Sulphur dioxide has a choking acidic smell. Its identification can be confirmed by either:

- Bubbling through a solution of potassium manganate(VII). The solution turns from purple to colourless.
- Soaking a piece of filter paper in potassium dichromate solution. The 'dichromate paper' is then placed over the mouth of the test tube. If sulphur dioxide is present, the paper turns from orange to green-blue.

Testing for halides

The test involves either aqueous silver nitrate or aqueous lead nitrate.

Testing with silver nitrate

The procedure is given in full in Section 12.8. Nitric acid is added to the suspected halide. Aqueous silver nitrate is then added followed by testing to see whether the precipitate dissolves in aqueous ammonia.

- Chlorides give a white precipitate of silver chloride:

  $$Ag^+(aq) + Cl^-(aq) \rightarrow AgCl(s)$$

  This precipitate dissolves readily in aqueous ammonia due to the formation of a complex ion:

  $$AgCl(s) + 2NH_3(aq) \rightarrow [Ag(NH_3)_2]^+(aq) + Cl^-(aq)$$

- Bromides give a cream precipitate of silver bromide:

  $$Ag^+(aq) + Br^-(aq) \rightarrow AgBr(s)$$

  The silver bromide precipitate dissolves only in excess aqueous ammonia.

- Iodides give a pale yellow precipitate of silver iodide which is insoluble in aqueous ammonia:

  $$Ag^+(aq) + I^-(aq) \rightarrow AgI(s)$$

Testing with lead nitrate

Nitric acid is added to the suspected nitrate. A solution of lead nitrate is then added.

- Chlorides give a white precipitate of lead(II) chloride:

  $$Pb^{2+}(aq) + 2Cl^-(aq) \rightarrow PbCl_2(s)$$

- Bromides give a pale yellow precipitate of lead(II) bromide.
- Iodides give a deep yellow precipitate of lead(II) iodide.

The addition of iodide ions to a solution of lead ions can also act as a confirmatory test for Pb^{2+} ions.

Testing for chromates

The chromate ion has the formula, CrO_4^{2-}. We add a few cubic centimetres of dilute sulphuric acid to the solution of the suspected chromate. If the chromate ion is present, the yellow colour of the chromate ions turns orange due to the formation of dichromate ions:

$$\underset{\text{chromate}}{2CrO_4^{2-}(aq)} + 2H^+(aq) \rightarrow \underset{\text{dichromate}}{Cr_2O_7^{2-}(aq)} + H_2O(l)$$

When a drop of dilute hydrogen peroxide is added to the dichromate, a blue colouration is seen which quickly turns green.

Key points

Ions can be identified using the following aqueous solutions:

- CO_3^{2-} using hydrochloric acid and then limewater
- NO_3^- using Cu and H_2SO_4 or Al and OH^-
- SO_4^{2-} using $Ba(NO_3)_2$
- SO_3^{2-} using H_2SO_4
- Halides using $AgNO_3$ or $Pb(NO_3)_2$
- CrO_4^{2-} using acid and H_2O_2

Exam-style questions Module 3

Answers to all exam-style questions can be found on the accompanying CD

Multiple-choice questions

1 Which of the following are properties of ceramic material?

i chemically inert
ii very high melting point
iii high tensile strength
iv resists compression

A i, ii
B i, ii, iii
C ii, iii, iv
D i, ii, iv

2 Which represents the correct order of increasing ionic radius?

A Cl^-, P^{3-}, Al^{3+}, Na^+
B P^{3-}, Cl^-, Na^+, Al^{3+}
C Al^{3+}, Na^+, Cl^-, P^{3-}
D Na^+, Al^{3+}, P^{3-}, Cl^-

3 A chloride in acid solution:

i gives a white or grey precipitate with aqueous mercury(II) chloride
ii decolourises aqueous potassium manganate(VII).

Which of the following chlorides will undergo the above reactions?

A $PbCl_4$
B $SnCl_2$
C $PbCl_2$
D $SnCl_4$

4 Which of the statements below is correct?

A Carbon monoxide turns blue litmus pink
B Lead(IV) chloride is more stable to heat than silicon(IV) chloride.
C The chemistry of germanium is significantly affected by the inert-pair effect.
D The screening effect of the inner electrons in lead is not very effective.

5 An element exhibits the following properties:

i The chloride is easily hydrolysed.
ii It accepts lone pairs forming co-ordination compounds.
iii It does not react at all readily with water.

Which element best fits the above properties?

A Be
B Mg
C Sr
D Ba

Structured questions

6 a i Define the term 'transition metal'. [2]
ii State four characteristics of a transition metal. [4]
iii Write the electronic configuration for the Fe^{3+} and Mn^{3+} ions, showing only the occupation of electrons in the orbitals of the valence shell. [2]

b Use the information provided in part **iii** to explain the following observations:
i Fe^{2+} ions are easily oxidised to Fe^{3+} ions. [2]
ii Mn^{2+} ions resist oxidation to Mn^{3+} ions. [2]

c When iron(III) chloride is dissolved in water a red-brown precipitate is formed.
i Give an explanation for the above observation. [2]
ii Write an equation to represent the reaction. [1]

7 a Define the terms
i co-ordination complex [2]
ii ligand [2]

b When concentrated hydrochloric acid is slowly added to a pale blue solution of copper(II) sulphate until present in excess, a yellow solution is produced.
i Write the formulae of the species responsible for the pale blue and yellow solutions respectively. [2]
ii Write the equation for the reaction. [2]
iii Using the concept of stability constant, carefully explain the above colour change. [3]

c i Describe what would be observed when aqueous ammonia is slowly added until excess to an aqueous solution of copper(II) sulphate. [3]
ii Write the formula for the final copper species formed. [1]

8 a Explain the difference in the reaction (if any) of CCl_4 and $SiCl_4$ with water and write equations for any reactions involved. [5]

b The standard electrode potential for the Ge^{4+}(aq)/Ge^{2+}(aq) system is –1.6 V. Use similar values for Sn and Pb to deduce the relative stability of the +2 and +4 oxidation states of the Group IV elements. [3]

c Chlorine disproportionates when bubbled into a hot concentrated solution of potassium hydroxide.

i Write an ionic equation to show what happens and state the oxidation number of the chlorine species involved. [2]

ii Write two ionic half equations to describe the process and thereby explain the term 'disproportionation'. [5]

9 a Give explanations for the following statements:

i $BaSO_4$ is less soluble than $MgSO_4$. [2]

ii Magnesium carbonate decomposes at 350 °C while strontium carbonate's decomposition temperature is 1340 °C. [4]

b Explain the trends in

i atomic radius [2]

ii ionic radius [1]

of the Group II elements.

c A compound of a Group II element (M) on strong heating evolves a brown gas and a white metallic oxide, N, containing 10.46% oxygen by mass. N readily dissolves to produce a colourless aqueous solution, S.
S when reacted with an appropriate dilute acid produces M on crystallising from the resulting solution.

i State the formula of the brown gas. [1]

ii Deduce the atomic mass of the element M. [2]

iii State the name and formula of M. [1]

iv Write the equation which represents the formation of M from S. [2]

10

Element/symbol	**b.pt.**	**density**
F		
Cl		
Br		
I		

a i Copy the above table and indicate by the use of vertical arrows the trends of the above physical properties of the elements as the group is descended. [2]

ii Explain your answers to part **i** in terms of structure and bonding with regards to b.pt. and density. [4]

b A white crystalline solid A when treated with concentrated sulphuric acid readily evolves a white fuming gas B as well as a reddish vapour. B when dissolved in water forms a solution C which produces a colourless gas, D, with sodium carbonate, which when bubbled through a solution of calcium hydroxide produces a white precipitate.

When aqueous silver nitrate is added to C a cream precipitate forms which dissolves in concentrated aqueous ammonia.

i State the names of the two gases B and D. [2]

ii State the formulae of the species responsible for the production of the gases.
Identifying the observations underlying the above answers. [2]

iii Write two ionic equations to explain the observations associated with the reaction of C with silver nitrate. [2]

c Use relevant information involving electrode potentials to explain the relative oxidising power of the halogens. [3]

Data sheets

Selected bond energies

Diatomic molecules		Polyatomic molecules	
Bond	**Bond energy/kJ mol^{-1}**	**Bond**	**Bond energy/kJ mol^{-1}**
H–H	436	C–C	350
N≡N	994	C=C	610
O=O	496	C–H	410
F–F	158	C–Cl	340
Cl–Cl	244	C–Br	280
Br–Br	193	C–I	240
I–I	151	C–N	305
H–F	562	C–O	360
H–Cl	431	C=O	740
H–Br	366	N–H	390
H–I	299	N–N	160
		O–H	460
		O–O	150

Selected electrode potentials

Electrode reaction	$E^{\ominus}$/V
$K^+ + e^- \rightleftharpoons K$	–2.92
$Mg^{2+} + 2e^- \rightleftharpoons Mg$	–2.38
$Al^{3+} + 3e^- \rightleftharpoons Al$	–1.66
$V^{2+} + 2e^- \rightleftharpoons V$	–1.2
$Zn^{2+} + 2e^- \rightleftharpoons Zn$	–0.76
$Fe^{2+} + 2e^- \rightleftharpoons Fe$	–0.44
$V^{3+} + e^- \rightleftharpoons V^{2+}$	–0.26
$Ni^{2+} + 2e^- \rightleftharpoons Ni$	–0.25
$Sn^{2+} + 2e^- \rightleftharpoons Sn$	–0.14
$Pb^{2+} + 2e^- \rightleftharpoons Pb$	–0.13
$2H^+ + 2e^- \rightleftharpoons H_2$	0.00
$S_4O_6^{2-} + 2e^- \rightleftharpoons 2S_2O_3^{2-}$	+0.09
$Cu^{2+} + 2e^- \rightleftharpoons Cu$	+0.34
$VO^{2+} + 2H^+ + e^- \rightleftharpoons V^{3+} + H_2O$	+0.34
$I_2 + 2e^- \rightleftharpoons 2I^-$	+0.54
$Fe^{3+} + e^- \rightleftharpoons Fe^{2+}$	+0.77
$Ag^+ + e^- \rightleftharpoons Ag$	+0.80
$VO_2^+ + 2H^+ + e^- \rightleftharpoons VO^{2+} + H_2O$	+1.00
$Br_2 + 2e^- \rightleftharpoons 2Br^-$	+1.07
$Cr_2O_7^{2-} + 14H^+ + 6e^- \rightleftharpoons 2Cr^{3+} + 7H_2O$	+1.33
$Cl_2 + 2e^- \rightleftharpoons 2Cl^-$	+1.36
$MnO_4^- + 8H^+ + 5e^- \rightleftharpoons Mn^{2+} + 4H_2O$	+1.52

Key

atomic (proton) number **atomic symbol** name relative atomic mass

IA	IIA	IIIB	IVB	VB	VIB	VIIB	VIIIB	VIIIB	VIIIB	IB	IIB	IIIA	IVA	VA	VIA	VIIA	VIIIA
							1 **H** hydrogen 1.008										2 **He** helium 4.003
3 **Li** lithium 6.941	4 **Be** beryllium 9.012											5 **B** boron 10.81	6 **C** carbon 12.01	7 **N** nitrogen 14.01	8 **O** oxygen 16.00	9 **F** fluorine 19.00	10 **Ne** neon 20.18
11 **Na** sodium 22.99	12 **Mg** magnesium 24.31											13 **Al** aluminium 26.98	14 **Si** silicon 28.09	15 **P** phosphorus 30.97	16 **S** sulfur 32.07	17 **Cl** chlorine 35.45	18 **Ar** argon 39.95
19 **K** potassium 39.10	20 **Ca** calcium 40.08	21 **Sc** scandium 44.96	22 **Ti** titanium 47.87	23 **V** vanadium 50.94	24 **Cr** chromium 52.00	25 **Mn** manganese 54.94	26 **Fe** iron 55.85	27 **Co** cobalt 58.93	28 **Ni** nickel 58.69	29 **Cu** copper 63.55	30 **Zn** zinc 65.39	31 **Ga** gallium 69.72	32 **Ge** germanium 72.61	33 **As** arsenic 74.92	34 **Se** selenium 78.96	35 **Br** bromine 79.90	36 **Kr** krypton 83.80
37 **Rb** rubidium 85.47	38 **Sr** strontium 87.62	39 **Y** yttrium 88.91	40 **Zr** zirconium 91.22	41 **Nb** niobium 92.91	42 **Mo** molybdenum 95.94	43 **Tc** technetium [98]	44 **Ru** ruthenium 101.1	45 **Rh** rhodium 102.9	46 **Pd** palladium 106.4	47 **Ag** silver 107.9	48 **Cd** cadmium 112.4	49 **In** indium 114.8	50 **Sn** tin 118.7	51 **Sb** antimony 121.8	52 **Te** tellurium 127.6	53 **I** iodine 126.9	54 **Xe** xenon 131.3
55 **Cs** caesium 132.9	56 **Ba** barium 137.3	57 **La** lanthanum 138.9	72 **Hf** hafnium 178.5	73 **Ta** tantalum 180.9	74 **W** tungsten 183.8	75 **Re** rhenium 186.2	76 **Os** osmium 190.2	77 **Ir** iridium 192.2	78 **Pt** platinum 195.1	79 **Au** gold 197.0	80 **Hg** mercury 200.6	81 **Tl** thallium 204.4	82 **Pb** lead 207.2	83 **Bi** bismuth 209.0	84 **Po** polonium [209]	85 **At** astatine [210]	86 **Rn** radon [222]
87 **Fr** francium [223]	88 **Ra** radium [226]	89 **Ac** actinium [227]	104 **Rf** rutherfordium [261]	105 **Db** dubnium [262]	106 **Sg** seaborgium [266]	107 **Bh** bohrium [264]	108 **Hs** hassium [269]	109 **Mt** meitnerium [268]									

58 **Ce** cerium 140.1	59 **Pr** praseodymium 140.9	60 **Nd** neodymium 144.2	61 **Pm** promethium [145]	62 **Sm** samarium 150.4	63 **Eu** europium 152.0	64 **Gd** gadolinium 157.3	65 **Tb** terbium 158.9	66 **Dy** dysprosium 162.5	67 **Ho** holmium 164.9	68 **Er** erbium 167.3	69 **Tm** thulium 168.9	70 **Yb** ytterbium 173.0	71 **Lu** lutetium 175.0
90 **Th** thorium 232.0	91 **Pa** protactinium [231]	92 **U** uranium 238.0	93 **Np** neptunium [237]	94 **Pu** plutonium [244]	95 **Am** americium [243]	96 **Cm** curium [247]	97 **Bk** berkelium [247]	98 **Cf** californium [251]	99 **Es** einsteinium [252]	100 **Fm** fermium [257]	101 **Md** mendelevium [258]	102 **No** nobelium [259]	103 **Lr** lawrencium [262]

Answers to revision questions

Module 1

Chapter 1 (page 12)

1 All atoms of the same element are exactly alike, atoms cannot be broken down any further, atoms of different elements have different masses, atoms combine to form more complex structures (compounds).

Atoms are not indestructible. Atoms are split in nuclear reactions, and they are made up of even smaller particles (subatomic particles).

Atoms of the same element can have different masses (not all are identical), that is, they possess isotopes.

2 Electrons are deflected to the positive plate because they are negatively charged, protons are deflected to the negative plate because they are positively charged and neutrons are not deflected at all because they have no charge.
The degree of deflection of an electron is more than that of a proton because an electron is lighter in mass.

3 The absolute masses and charges of sub-atomic particles are very small, so it is much easier to use relative masses and charges when making comparisons and doing calculations.

4 a 13 protons, 13 electrons, 14 neutrons
b 19 protons, 19 electrons, 20 neutrons
c 53 protons, 53 electrons, 78 neutrons
d 94 protons, 94 electrons, 145 neutrons

5 $\frac{(10 \times 18.7) + (11 \times 81.3)}{100} = 10.8$

6 a i $^{238}_{92}U \rightarrow {}^{234}_{90}Th + {}^{4}_{2}He$ ii $^{222}_{88}Ra \rightarrow {}^{218}_{86}Rn + {}^{4}_{2}He$
b i $^{234}_{90}Th \rightarrow {}^{234}_{91}Pa + {}^{0}_{-1}e$ ii $^{14}_{6}C \rightarrow {}^{14}_{7}N + {}^{0}_{-1}e$

7 Iodine-131 is used to study thyroid function
Carbon-14 is used for dating objects which were once living
Americium-241 is used in smoke detectors
Uranium-235 is used to generate energy in nuclear reactors

8 Electrons are arranged in energy levels which are quantised.
When an electron absorbs a quantum of energy of a particular frequency or wavelength it becomes excited and moves up to a higher energy level.
The electron eventually loses this quantum of energy and moves back down to a lower energy level.
As it does so, the energy that it loses is given out and is seen as a line of that particular frequency or wavelength in the emission spectrum.

9 a Previously excited electrons fall back to the $n = 1$ energy level.
b Previously excited electrons fall back to the $n = 2$ energy level.

10 $\Delta E = h\nu$
ΔE is the energy difference between the two energy levels.
h is Planck's constant.
ν is the frequency of the radiation.

11 2, 6 and 10

12 a A region of space where there is a high probability of finding an electron
b

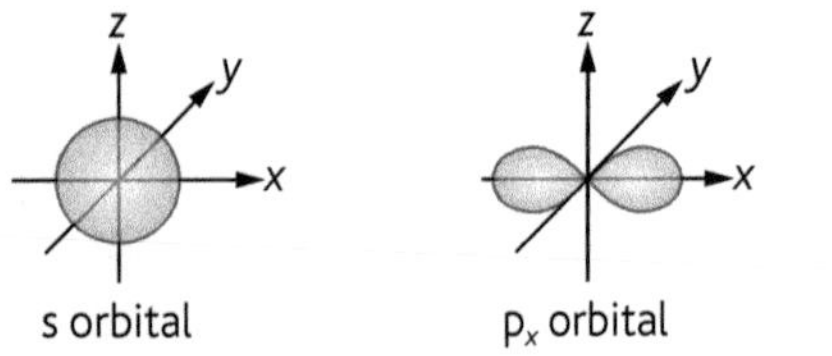

13 a $1s^2 2s^2 2p^6 3s^2 3p^6 4s^2$
b $1s^2 2s^2 2p^6 3s^2 3p^6 3d^{10} 4s^2$
c $1s^2 2s^2 2p^6 3s^2 3p^6$
d $1s^2 2s^2 2p^6 3s^2 3p^6 3d^5$

14 a The first ionisation energy, ΔH_{i1}, is the energy needed to remove one electron from each atom in one mole of atoms of an element in its gaseous state to form one mole of gaseous ions.
b $Ca(g) \rightarrow Ca^{+}(g) + e^{-}$

15 Size of nuclear charge, distance of the outer electrons from the nucleus, shielding

16 This is due to the increase in nuclear charge as one goes across the period.

17 a The electron in Al is removed from the higher energy 3p orbital compared to Mg where the electron to be removed comes from the lower energy 3s orbital. It therefore requires less energy to remove the electron in Al, therefore it has a lower first ionisation energy.
b The electron to be removed from S comes from the $3p_x$ orbital where it is spin-paired with another electron. The repulsion between these two electrons means that this electron is easier to remove than the electron in P, which is also in the 3p orbital but is not spin-paired.

18 Group II

Chapter 2 (page 32)

1 a i Sigma bonds (σ bonds) are formed by the end-on overlap of atomic orbitals.
ii Pi bonds (π bonds) are formed by the sideways overlap of p atomic orbitals.
b Sigma bonds are stronger because the electron density is symmetrical about a line joining the two nuclei, whereas in a pi bond the electron density is not symmetrical.

2 a

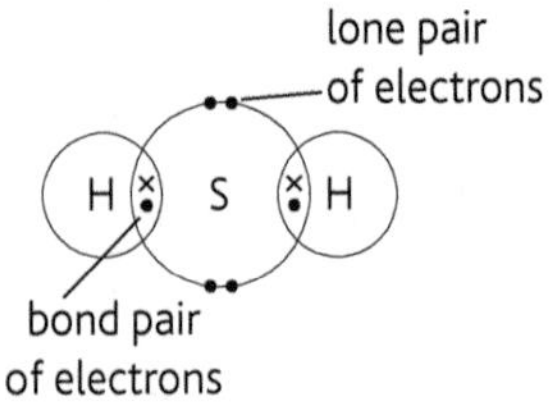

b

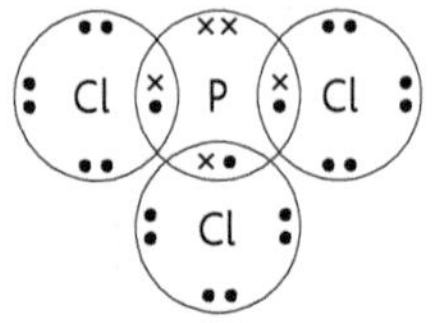

c

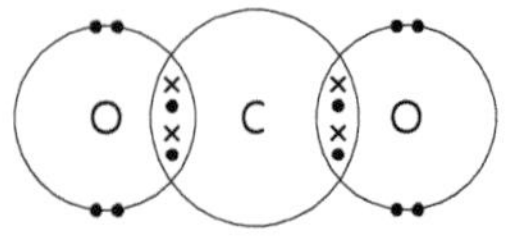

3 a A co-ordinate bond is formed when one atom provides both the electrons for the covalent bond.
b

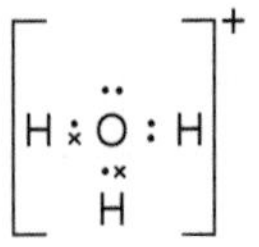

4 **a** **b**

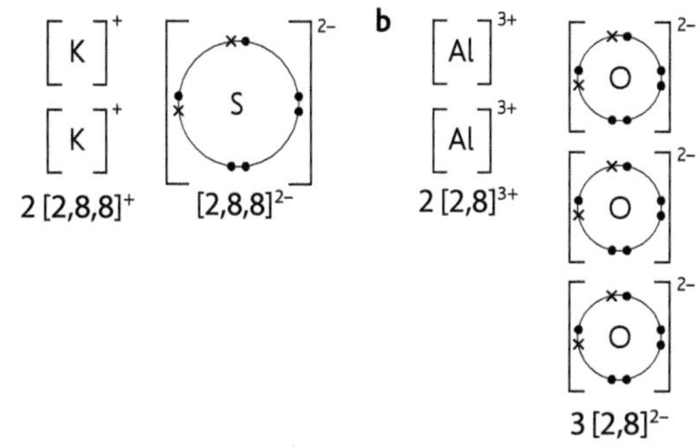

5 **a** Because the element has one electron in its outer shell it is a metal, therefore the bonding is metallic.
In this type of bonding a lattice of metal ions is surrounded by a 'sea' of delocalised electrons.

b The melting point would be high.
This strong electrostatic attraction between the metal ions and the delocalised electrons acts in all directions and results in the bond being strong and hard to break. As a result the melting point will be high.

c sodium, magnesium, aluminium

6 **a** Electronegativity is the ability of a particular atom involved in covalent bond formation to attract the bonding pair of electrons towards itself.

b Electronegativity decreases.

c As you go down Group VII, electronegativity decreases because the size of the atoms increase resulting in the bonding pair of electrons being increasingly further from the attraction of the nucleus.

7 **a** non-polar **c** non-polar
b polar **d** polar

8 **a** Permanent dipole–dipole forces, van der Waals forces, hydrogen bonding

b **i** van der Waals forces
ii hydrogen bonding
iii hydrogen bonding
iv permanent dipole–dipole forces

9 Magnesium chloride has a giant ionic structure, diamond has a giant covalent structure and aluminium is metallic.

a All three solids have high melting points. In magnesium chloride, strong ionic bonds due to the electrostatic forces of attraction between the oppositely charged ions must be broken. In diamond, strong covalent bonds between the carbon atoms must be broken and, in aluminium, strong metallic bonds due to the forces of attraction between the metal ions and the delocalised electrons must be broken.

b Magnesium chloride will conduct electricity when molten or dissolved in water because the ions which make up the structure are now free to move.
Diamond cannot conduct electricity in any state due to the fact that it possesses no ions or electrons that are free to move.
Aluminium will conduct electricity in both the solid and molten state because the delocalised electrons are free to move.

c Magnesium chloride is soluble in water due to the fact that the ions present can form ion–dipole bonds with water molecules.
Diamond is not soluble in water because the atoms are held together to form bonds with water.
Aluminium is not soluble in water because the force of attraction between the metal ions and the delocalised electrons is too strong to allow the ions to form bonds with water.

10 Iodine is a simple molecular solid:

a Weak van der Waals forces exist between the molecules, allowing the solid to break easily.

b There are no free ions or delocalised electrons.

c Because the forces of attraction between the molecules are so weak that not much energy is needed to turn the solid into a liquid and the liquid into a vapour.

11 Water has a much higher boiling point due to the fact that it is extensively hydrogen bonded unlike the other hydrides which only possess van der Waals forces between their molecules.

12 PCl_3 would have a higher melting point than BCl_3 because of the greater number of electrons it possesses, which therefore produce stronger van der Waals forces which require more energy to break. Also it can be said that the mass of PCl_3 is greater than that of BCl_3, again causing an increase in the strength of its van der Waals forces.

13 **a** Pyramidal – there are three bond pairs and one lone pair on the oxygen atom.

b Non-linear – there are two bond pairs and two lone pairs on the sulphur atom.

c Trigonal planar – there are three bond pairs.

d Tetrahedral – there are four bond pairs.

e Pyramidal – there are three bond pairs and one lone pair on the phosphorus atom.

Chapter 3 (page 46)

1 **a** $FeCl_3(aq) + 3NaOH(aq) \rightarrow Fe(OH)_3(s) + 3NaCl(aq)$
$Fe^{3+}(aq) + 3OH^-(aq) \rightarrow Fe(OH)_3(s)$

b $Mg(s) + 2HCl(aq) \rightarrow MgCl_2(aq) + H_2(g)$
$Mg(s) + 2H^+(aq) \rightarrow Mg^{2+}(aq) + H_2(g)$

c $Na_2S_2O_3(aq) + 2HCl(aq) \rightarrow 2NaCl(aq) + H_2O(l) + S(s) + SO_2(g)$
$S_2O_3^{2-}(aq) + 2H^+(aq) \rightarrow H_2O(l) + S(s) + SO_2(g)$

2 **a** **i** A mole is the amount of substance which has the same number of specified particles as there are atoms in exactly 12 g of the carbon-12 isotope.
ii Molar mass is the mass of 1 mole of a substance in grams.

b **i** $331\,g\,mol^{-1}$ **ii** $310\,g\,mol^{-1}$
iii $250\,g\,mol^{-1}$

3 **a** **i** $7.1/142 = 0.05\,mol$ **ii** $20/100 = 0.2\,mol$

b **i** $0.025 \times 84 = 2.1\,g$
ii $2.0 \times 10^{-3} \times 36.5 = 0.073\,g$

4 Balanced equation for reaction:
$C_3H_8(g) + 5O_2(g) \rightarrow 3CO_2(g) + 4H_2O(g)$
Number of moles of $C_3H_8 = 5.5/44 = 0.125\,mol$
From balanced equation, 1 mole C_3H_8 reacts with 5 moles of O_2.
So number of moles of $O_2 = 0.125 \times 5 = 0.625\,mol$
Mass of $O_2 = 0.625 \times 32 = 20\,g$

5 **a** Molecular formula is N_2O_4.
Empirical formula is NO_2.

b Molecular formula is S_2Cl_2.
Empirical formula is SCl.

6

	C	H
Number of moles	$\frac{0.48}{12} = 0.04$	$\frac{0.08}{1} = 0.08$
Mole ratio	$\frac{0.04}{0.04} = 1$	$\frac{0.08}{0.04} = 2$

Therefore empirical formula is CH_2.
Empirical formula mass is $12 + 2 = 14$

Divide molar mass by empirical formula mass = $\frac{56}{14}$ = 4
Multiply each atom in empirical formula by 4.
Therefore molecular formula is C_4H_8.

7 **a** Number of moles = 6/24 = 0.25 mol
Mass = 0.25 × 32 = 8 g
b Converting 120 cm^3 to dm^3
120/1000 = 0.12 dm^3
Number of moles at r.t.p. = 0.12/24 = 0.005 mol
Mass = 0.005 × 64 = 0.32 g

8 Volume of 56 cm^3 of butane in dm^3 = 56/1000 = 0.056 dm^3
Number of moles of butane at s.t.p. = 0.056/22.4 = 0.0025 mol
Balanced equation for reaction:

$$2C_4H_{10}(g) + 13O_2(g) \rightarrow 8CO_2(g) + 10H_2O(g)$$

From balanced equation, 2 moles C_4H_{10} react to produce 10 moles of H_2O.
Therefore 0.0025 moles C_4H_{10} produce $\frac{10}{2}$ × 0.0025 = 0.0125 mole of H_2O
Volume of steam at s.t.p. = 0.0125 × 22.4 = 0.28 dm^3

9 $C_xH_y(g) + \; O_2(g) \rightarrow xCO_2(g) + yH_2O(g)$
20 cm^3 160 cm^3 100 cm^3
1 vol 8 vol 5 vol
$C_5H_y(g) + 8O_2(g) \rightarrow 5CO_2(g) + yH_2O(g)$
10 moles of O reacts with C so the other 6 moles must react with H,
so 6 moles H_2O are formed which contain 12 atoms of H.
Therefore pentane must be C_5H_{12}.

10 **a** Number of moles of KNO_3 = $\frac{3.33}{101}$ = 0.033 mol
Converting 250 cm^3 to dm^3 = $\frac{250}{1000}$ = 0.25 dm^3
Molar concentration = $\frac{0.033}{0.25}$ = 0.13 mol dm^{-3}
b Number of moles of Na_2CO_3 = $\frac{2.65}{106}$ = 0.025 mol
Converting 75 cm^3 to dm^3: $\frac{75}{1000}$ = 0.075 dm^3
Molar concentration = $\frac{0.025}{0.075}$ = 0.33 mol dm^{-3}

11 **a** Change 25 cm^3 to dm^3: 25/1000 = 0.025 dm^3
Number of moles: 0.01 × 0.025 = 0.00025 mol
Mass: 0.00025 × 58.5 = 0.015 g
b Change 750 cm^3 to dm^3: 750/1000 = 0.75 dm^3
Number of moles: 0.27 × 0.75 = 0.20 mol
Mass: 0.20 × 80 = 16.0 g

12 **a**

Burette readings/ cm^3	1	2	3	4
Final volume	**27.15**	25.80	37.90	**26.50**
Initial volume	0.50	**0.15**	12.50	1.00
Actual volume	26.65	25.65	**25.40**	25.50

b Average volume of acid = $\frac{(25.40 + 25.50)}{2}$
= 25.45 cm^3

13 Number of moles of potassium manganate(VII) used:

$$0.2 \times \frac{27.50}{1000} = 0.0055 \text{ mol}$$

From balanced equation:

$$MnO_4^-(aq) + 5Fe^{2+}(aq) + 8H^+(aq) \rightarrow Mn^{2+}(aq) + 5Fe^{3+}(aq) + 4H_2O(l)$$

1 mol of potassium manganate(VII) reacts with 5 mol iron(II)
Therefore 0.0055 mol potassium manganate(VII) reacts with 5 × 0.0055 = 0.0275 mol Fe^{2+}
This is the number of moles in 25.0 cm^3 of iron(II).
Converting 25.0 cm^3 to dm^3: 25.0/1000 = 0.025 dm^3
Therefore molar concentration: 0.0275/0.025 = 1.10 mol dm^{-3}

Module 2

Chapter 7 (page 88)

1 **a** Set up apparatus similar to Figure 7.1.1, substitute the different reactants and measure the volume of N_2 at specific time intervals.
b The $I_2(aq)$ is brown in colour.

2 **a** I and III
b Determine at least the first two half-lives. For a first order reaction, the half-lives are constant. For a second order reaction, the half-lives become longer.
c The graph would be an upward curve.

3 **a** Second order
b First order
c Third order
d Rate = $k[X]^2[Y]$
e $k = \frac{\text{Rate}}{[X]^2[Y]}$ = 9200 $dm^6 mol^{-2} s^{-1}$

4 **a** Draw a tangent to the curve at the time given. [There may be variations in the answer depending on how the tangent is drawn.]
i For time = 0 s, Initial rate = 1.1×10^{-2} mol $dm^{-3} s^{-1}$
ii For time = 20 s, Rate = 7.6×10^{-3} mol $dm^{-3} s^{-1}$
For time = 60 s, Rate = 3.4×10^{-3} mol $dm^{-3} s^{-1}$
b i Half-life is the time taken for the initial concentration to decrease to half the value.
ii First $t_{½}$ = 35 s, Second $t_{½}$ = 35 s
c First order, half-life is constant.

5 **a** Kinetic energy (*x*-axis), fraction of particles (*y*-axis)
b Refer to Section 7.3.
c i Rate would decrease.
ii Refer to Section 7.3.

6 **a** D
b C
c C
d A

7 **a** i $NO(g) + SO_2(g) + O_2(g) \rightarrow SO_3(g) + NO_2(g)$
ii Step 1
iii Rate = $k[SO_2][NO_2]$, the NO_2 functions as a catalyst.
b i Mechanism I
ii The rate would double.

Chapters 8–10 (page 130)

1 **a** i $K_c = \frac{[NO_2(g)]^2[Cl_2(g)]}{[NO_2Cl(g)]^2}$ Units: mol dm^{-3}
ii $K_c = \frac{[Fe^{2+}(aq)][I_2(aq)]^{½}}{[Fe^{3+}(aq)][I^-(aq)]}$ Units: $mol^{½}$ $dm^{1½}$
b i $K_p = \frac{(p_{O_3})^2}{(p_{O_2})^3}$ Units: atm^{-1}
ii $K_p = \frac{(p_{HF})^3 (p_{N_2})^{½} (p_{Cl_2})^{½}}{(p_{NH_3}) (p_{ClF_3})}$ Units: atm^2
c i $K_{sp} = [IO_3^-(aq)]^2[Ba^{2+}(aq)]$ Units: $mol^3 dm^{-9}$
ii $K_{sp} = [Al^{3+}(aq)][OH^-(aq)]^3$ Units: $mol^4 dm^{-12}$

d i $K_a = \frac{[H_3O^+(aq)][F^-(aq)]}{[HF(aq)]}$ Units: $mol\,dm^{-3}$

ii $K_a = \frac{[H^+(aq)][CN^-(aq)]}{[HCN(aq)]}$ Units: $mol\,dm^{-3}$

e $K_b = \frac{[H_2S(aq)][OH^-(aq)]}{[HS^-(aq)]}$ Units: $mol\,dm^{-3}$

2 a Dynamic equilibrium is the state that exists for a closed system, where the rate of the forward reaction or process is equal to the rate of the reverse reaction or process. The concentration of products and reactants at equilibrium remain constant and the system can be represented by an equilibrium constant *K*. The value of *K* is only affected by changing the temperature of the system. An example of a system in dynamic equilibrium is $N_2(g) + 3H_2(g) \rightleftharpoons 2NH_3(g)$.

b

	Yield	Position of equilibrium	Value of equilibrium constant
i	decrease	shift to left	no effect
ii	increase	shift to right	no effect
iii	no effect	no effect	no effect
iv	increase	shift to right	increase
v	increase	shift to right	no effect

c i No effect on position of equilibrium, yield or value of equilibrium constant

ii Yield increases, shift in position to right, no effect on value of constant

iii Yield decreases, shift to the left, value of equilibrium constant decreases

3 a $K_c = 15.3\,mol\,dm^{-3}$

b $[R] = 0.14\ mol\,dm^{-3}$

c $K_c = 33.4$

d $K_c = 0.53\,mol\,dm^{-3}$

4 a $p_{H_2} = 3.7\,atm$, $p_{N_2} = 0.2\,atm$

b $K_p = 5.0 \times 10^{-4}\ Pa^{-1}$

c $K_p = 0.24$

5 a The maximum amount of a solute that dissolves in a given volume of a solvent at a particular temperature

b i $K_{sp} = 5.67 \times 10^{-5}\ mol^2\ dm^{-6}$

ii $K_{sp} = 1.70 \times 10^{-6}\ mol^3\ dm^{-9}$

c i For pure water, solubility is $1.9 \times 10^{-4}\,mol\,dm^{-3}$. For a solution of $0.015\,mol\,dm^{-3}\ Mg^{2+}$ ions, solubility is $2.4 \times 10^{-6}\,mol\,dm^{-3}$.

ii The common ion effect

iii $MgCO_3$ would precipitate, $[Mg^{2+}][CO_3^{2-}] > K_{sp}$

6 a i HCl is a strong acid and dissociates completely, whereas C_6H_5COOH is a weak acid and dissociates partially.

ii pH = 1.6

iii pH = 2.9

b pH = 1.7, pOH = 12.3

c i $[H^+] = 3.98 \times 10^{-12}\ mol\,dm^{-3}$

ii $[OH^-] = 2.51 \times 10^{-3}\ mol\,dm^{-3}$

7 a i $C_6H_5NH_2(aq) + H_2O(l) \rightleftharpoons C_6H_5NH_3^+(aq) + OH^-(aq)$

ii $C_6H_5NH_3^+$

iii pH = 9.0, $pK_b = 9.4$

b i The weak acid $HC_3H_5O_3$ would neutralise any base added and its conjugate base. $C_3H_5O_3^-$ from the salt would neutralise any acid added. Nitric acid is a strong acid and the NO_3^-, from the salt does not function as a base.

ii K_a of $CH_3COOH = 1.8 \times 10^{-5}\,mol\,dm^{-3}$
$[CH_3COOH] = 1.03\,mol\,dm^{-3}$,
$[CH_3COO^-] = 0.756\,mol\,dm^{-3}$,
$\therefore pH = 4.61$

iii K_b of $CO_3^{2-} = 2.1 \times 10^{-4}\,mol\,dm^{-3}$
pH = 10.3

8 a pH = 13.2

b pH = 12.7

c pH = 1.9

9 a Diagram similar to Figure 10.2.2, with the $Sn^{4+}(aq)/Sn^{2+}(aq)$ connected to the standard hydrogen electrode.

b i $Al(s)\,|\,Al^{3+}(aq)\,||\,Zn^{2+}(aq)\,|\,Zn(s)$

ii $E^{\ominus}_{cell} = +0.90\,V$

iii $2Al(s) + 3Zn^{2+}(aq) \rightarrow 2Al^{3+}(aq) + 3Zn(s)$

Module 3

Chapters 11–12 (page 156)

1 A 2 C 3 C 4 C 5 B

6 a Monodentate ligand can only form one bond to the central atom, e.g., ammonia molecule, $H_3N\!:\!\rightarrow$ bidentate ligand can form two bonds to the central atom; e.g., ethane-1,2-diamine, $H_2NCH_2CH_2NH_2$ (↓ ↓)

b i tetrachlorocuprate(II) and hexaaquairon(III)

ii
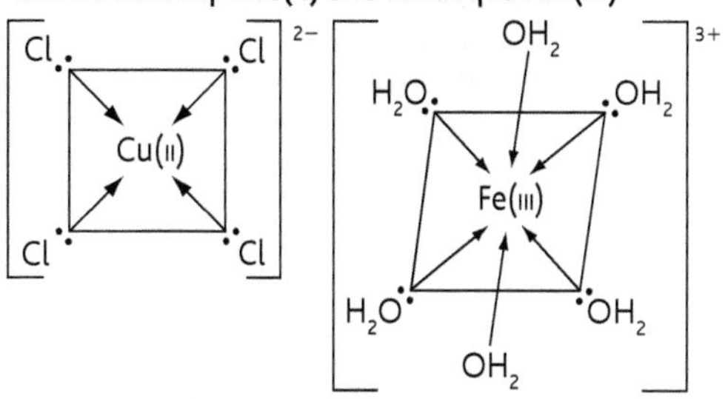

c i $[Cu(H_2O)_6]^{2+}(aq) + 4Cl^-(aq) \rightarrow CuCl_4^{2-}(aq) + 6H_2O(l)$

ii Co-ordination no. 6; co-ordination no. 4

iii The chlorine ion is much larger than the water molecule and the copper atom can therefore only accommodate four of these ions.

7 a In Period 3, oxidation states rise steadily as the elements either lose or gain all their valence electrons. In case of electropositive elements, Na–Al, positive oxidation states rise from +1 to +3; from silicon to chlorine electrons are shared in covalent bonds rising from +4 to +7.

e.g.	Na_2O	Al_2O_3	P_2O_5	Cl_2O_7
oxidation number	+1	+3	+5	+7

b For NaCl and $MgCl_2$: structure ionic crystal lattice. Bonding – strong ionic (electrovalent) bonds, ions held together by electrostatic forces.
On addition of water the crystal lattice is destroyed resulting in a neutral solution containing hydrated ions with a pH of 7.
For aluminium chloride: solid lattice structure. Bonding – ionic with appreciable covalent character. On addition of water the small and highly charged aluminium ion (large charge density) which forms the hexaaquaaluminium(III) ion, which weakens the O–H bond in the water molecules thus enabling protons to be extracted. This results in acid behaviour as exhibited by a pH of 5.
One of the possible equilibria is represented by the equation:

$$[Al(H_2O)_6]^{3+}(aq) + H_2O(l) \rightarrow [Al(H_2O)_5OH]^{2+}(aq) + H_3O^+(aq)$$

c Sodium reacts explosively with water producing hydrogen:

$$2Na(s) + 2H_2O(l) \rightarrow 2NaOH\ (aq) + H_2(g)$$

Aluminium is protected by an oxide film and does not react with water, however at elevated temperatures steam will convert the metal to its oxide:

$$2Al(s) + 3H_2O(g) \rightarrow Al_2O_3(s) + 3H_2(g)$$

Chlorine has a low solubility in water and forms a yellow-green solution, called chlorine water, which contains a mixture of hydrochloric and chloric(I) acids:

$$Cl_2(g) + H_2O(l) \rightarrow HCl\ (aq) + HClO\ (aq)$$

8 a The correct graph should be drawn from the given coordinates.

b i Rise in density from sodium to aluminium:
Structure: giant metallic structures
Bonding: metallic bonding – atomic radius decreases while charge increases resulting in stronger metallic bonds leading to close packing of atoms seen as a steady increase in density.

ii Decrease in density between sulphur and chlorine:
Structure: sulphur possesses a solid covalent structure. Chlorine exists as small, simple covalent gaseous molecules.
Bonding: sulphur's solid structure indicates stronger covalent bonding while only weak intermolecular van der Waals forces hold the gaseous chlorine molecules together. This results in the significant reduction in density.

9 a i All tetrahalides, MCl_4, exist as covalently bonded, simple tetrahedral molecules with weak van der Waals intermolecular bonds, hence their volatility. The symmetry of the tetrahedral structure leads to a net dipole moment of zero, thus their non-polar nature.

ii The reactivity of the tetrahalides increase as the group is descended – the elements change from non-metallic to metallic, and hence these compounds become more unstable, thus with water, fumes of hydrogen chloride gas are formed along with the oxide of oxidation state 4 (CCl_4 is exceptional, being unreactive).
$SiCl_4 + 2H_2O \rightarrow SiO_2 + 4HCl\ (g)$
$PbCl_4 + 2H_2O \rightarrow PbO_2 + 4HCl\ (g)$

b In carbon the + 4 oxidation state is more stable than the +2, thus CO is readily oxidised to CO_2. Hence CO is the better reducing agent.

c Silicon(IV) oxide:
Structure: macromolecule with each silicon atom surrounded by four others (diamond-like).
Bonding: Very strong covalent bonds with silicon atom in a tetrahedral environment.
The above properties lead to very high melting point; the absence of any unpaired electrons account for its insulating properties.

d

Oxides	Reactions	
	With acid	With alkali
Germanium(II) oxide	$Ge^{2+}(aq)$	$GeO_2^{\ 2-}(aq)$
Tin(II) oxide	$Sn^{2+}(aq)$	$SnO_2^{\ 2-}(aq)$
Lead(IV) oxide	$Pb^{2+}(aq)$	$PbO_3^{\ 2-}(aq)$

10 a As the group is descended, the decomposition temperatures of the carbonates increase.
Carbonates, MCO_3, decompose as this:

$$MCO_3(s) \rightarrow MO(s) + CO_2(g)$$

The following points are useful in explaining the trend:
The smaller size of the oxide ion, O^{2-}, allows it to be more strongly attached to the metal cation.
Hence making the crystal structure more thermally stable.
The resultant larger charge density of the oxide ion further strengthens the bonding to the metal ion.
The tendency for these carbonates to form the oxide will decrease down the group and therefore these carbonates become more resistant to decomposition as the group is descended.

Or

Large charge densities of cations allow for polarisation of the large anions. The charge density of the cations of Group II metals decrease as the group is descended, therefore their ability to polarise the larger carbonate ion to produce carbon dioxide and the oxide ion decreases. This results in the higher thermal stability of the carbonates lower down the group.

b i Solubility of the Group II sulphates decrease as the group is descended.

ii The solubility depends on the enthalpy change of solution, ΔH_{soln}.
By applying Hess's law ; $\Delta H_{soln} = \Delta H_{hyd} - \Delta H_{latt}$
Both ΔH_{hyd} and ΔH_{latt} decrease as the group is descended, however the decrease by ΔH_{hyd} is more significant than that of ΔH_{latt} (ΔH_{hyd} depends on the size of the cation). Therefore as the group is descended ΔH_{soln} becomes more positive resulting in decrease solubility of the sulphates.

c i calcium oxide: include reducing acidity of soils, paper manufacture, de-hairing hides, insecticides

ii magnesium oxide: laxative, medication (relieve acid indigestion), insulation for cables

iii calcium carbonate: cement, (concrete), glass, construction (road, housing), in the production of steel (blast furnace), reducing acidity in soils

d i Radium should react very vigorously with water, pH 13.

ii Radium chloride – soluble, radium carbonate – insoluble

iii Radium nitrate(V) would be most difficult to decompose.

11 a Melting points of Group VII increase as group is descended.
Structure: Changes from simple covalent molecules to covalent solids.
Bonding: Intermolecular forces become stronger as group is descended, – size of molecules increase.

b i $Br_2(l)$ added to KI (aq) – brown solution formed.
$Br_2(l)$ added to KCl (aq) – reddish brown colour.

ii Redox reaction

iii Iodine displaced by bromine, bromine oxidises iodide ions.
$2I^-(aq) + Br_2(l) \rightarrow I_2(aq) + 2Br^-(aq)$

c i P = MnO_2

ii Q = Cl_2

iii $MnO_2 + 4HCl \rightarrow MnCl_2 + Cl_2 + 2H_2O$

iv $Cl_2(g) + H_2O(l) \rightarrow HClO(aq) + HCl(aq)$

v Ox. state +1 −1

vi Dispropotionation

vii $2HClO(aq) \rightarrow 2HCl(aq) + O_2(g)$

Glossary

A

Acid A proton (hydrogen ion) donor.
Acid–base indicator A substance which changes colour when the pH in an acid–base titration changes rapidly.
Acid dissociation constant, K_a The equilibrium constant for a weak acid.
Activation energy The minimum energy particles must have when they collide to allow a reaction to occur.
Alkali A soluble base.
Allotrope Different forms of the same element. For example, diamond and graphite are allotropes of carbon.
Alpha radiation (α) The He^{2+} nuclei emitted when radioactive atoms decay.
Amphoteric oxide An oxide which can react with an acid as well as with an alkali.
Anion A negatively charged ion.
Atomic number (*Z*) The number of protons in the nucleus of the atom
Atomic orbital A region of space outside the nucleus where there is a good probability of finding one or two specific types of electron. Orbitals can be s, p, d or f.
Atomic radius/covalent radius Half the distance between the nuclei of two atoms of the same type in a covalent bond.
Avogadro constant The number of atoms, ions or molecules in a mole of atoms, ions or molecules.
Avogadro's law Equal volumes of gases contain the same number of molecules.

B

Base A proton (hydrogen ion) acceptor.
Beta radiation (β) Electrons given off from the nucleus when a radioactive isotope decays.
Bidentate Ligands which can form two co-ordinate bonds with the central transition element ion.
Boltzmann distribution curve A graph of the fraction of particles with particular energies against the energy. It has a characteristic humped curve which 'tails' off gradually.
Bond energy The energy needed to break a mole of a particular bond in one mole of gaseous molecules (under standard conditions).
Bond polarisation A covalent bond where the bonding electrons are not shared equally due to differences in electronegativity. Shown by an arrow and the signs δ^+ and δ^-.
Born–Haber cycle An enthalpy cycle used to calculate lattice energy.
Boyle's law The volume of a gas is inversely proportional to its pressure.
Brønsted–Lowry theory An acid is a proton donor. A base is a proton acceptor.
Buffer solution A solution that minimises changes in pH when acids or alkalis are added to the solution.

C

Catalyst A substance that increases the rate of reaction without change to itself.
Cation A positively charged ion.
Charles' law The volume of a gas is directly proportional to the temperature in kelvins.
Closed system A system in which matter or energy is not gained or lost.
Common ion effect The reduction in the solubility of a sparingly soluble salt by adding another salt which has an ion in common.
Complex ion An ion containing a central transition element bonded to ligands by co-ordinate bonds.
Compound A substance containing atoms from two or more different elements bonded together.
Condensation The change in state when a vapour changes to a liquid.
Conjugate pair An acid and base on each side of an equation which differ from each other by a H^+ ion.
Co-ordinate bond A covalent bond where both the electrons are provided by the same atom.
Co-ordination number The number of co-ordinate bonds formed by ligands to the central transition element ion in a complex.
Covalent bond (single) A bond formed by the sharing of a pair of electrons between two atoms.
Covalent radius See atomic radius.

D

Dative covalent bond See Co-ordinate bond.
Degenerate orbitals Atomic orbitals at the same energy level.
Delocalised electrons Electrons which are free to move between three or more adjacent atoms.
Diatomic Elements or compounds which have molecules containing only two atoms. For example, chlorine, carbon monoxide.
Dipole A separation of charge in a molecule. One end of the molecule has a partial positive charge and the other end a partial negative charge.
Displacement reaction A reaction where one atom or group of atoms replaces another in a chemical reaction.
Disproportionation A redox reaction in which the same type of atom in a species is oxidised and reduced simultaneously.
Dissociation The breaking up of a molecule into ions, especially referring to acids dissociating into H^+ and A^- ions.
d-orbital splitting The splitting of d-orbitals into a higher and lower energy level, caused by the presence of ligands around a central transition element ion.
Dot and cross diagram A diagram showing the arrangement of the outer shell electrons in a compound showing the origin of the electrons by dots or crosses.
Dynamic (equilibrium) In an equilibrium reaction, products are converted to reactants as well as reactants to products.

E

Electrical conductivity The ability of a substance to conduct electricity. In metals and graphite delocalized valence electrons are responsible for electrical conduction. In molten and aqueous salts, mobile ions are responsible for conduction.
Electrode potential See Standard electrode potential.
Electron affinity The first electron affinity is the enthalpy change when one mole of electrons is added to one mole of gaseous atoms to form one mole of gaseous 1+ ions.
Electronegativity The ability of a particular atom in a covalent bond to attract electrons to itself.
Electronic configuration The arrangement of electrons in shells and sub-shells e.g. $1s^22s^22p^5$.
Emission spectrum A series of bright lines seen in a spectroscope due to excited electrons falling from higher to lower energy levels.
Empirical formula The formula showing the simplest ratio of atoms in a molecule.
Endothermic A reaction which absorbs energy from the surroundings.

Energy level The regions at various distances from the nucleus where the electrons have defined amounts of energy.
Energy profile diagram Shows the enthalpy change from reactants to products along the reaction pathway.
Enthalpy change The energy transferred in a chemical reaction, ΔH.
Enthalpy cycle Shows alternative routes for a chemical reaction to proceed.
Enzyme A protein catalyst.
Equilibrium constant A constant, K_c or K_p, calculated from the equilibrium expression for a reaction.
Equilibrium expression A relationship linking K_c or K_p to the concentrations present at equilibrium.
Equilibrium reaction A reaction in which reactants and products are present in fixed concentrations. The rate of the forward and back reactions are equal.
Exothermic A reaction which gives out energy to the surroundings.

F

Feasibility The likelihood or not of a reaction occurring as predicted by E° values.
Ferromagnetic A substance which, when placed in a magnetic field, lines itself up with the field and retains its magnetism when the magnetic field is removed.

G

Gamma radiation (γ) Radiation given off an isotope decays and a proton is converted to a neutron by capturing an electron.
Gas constant The constant in the ideal gas equation, R.
Giant ionic structure A giant structure made up of positive and negative ions'.
Giant structure These have a three-dimensional network of covalent, metallic or ionic 'bonds'.

H

Half cell The half of an electrochemical cell, containing an element or ion in one oxidation state and an ion in another oxidation state.
Half equation An equation showing either the oxidation or reduction part of a redox equation balanced with electrons.
Half-life The time taken for the amount or concentration of a limiting reactant to decrease to half its initial value.
Halogens The Group VII elements.
Hess's law The enthalpy change for a reaction is independent of the route by which the reaction takes place.
Hybridisation The combination of atomic orbitals to form an orbital with mixed character.
Hydrogen bond A weak intermolecuar force formed between molecules with a H atom bonded to a F, O or N atom.
Hydrolysis The breakdown of a compound by reaction with water.

I

Ideal gas A gas which obeys Boyle's law and Charles' law under all conditions.
Ideal gas equation Relates volume, temperature, pressure and number of moles. $PV = nRT$.
Indicator See acid–base indicator and redox indicator.
Intermolecular forces Weak forces between molecules.
Ionic bond The electrostatic attraction between oppositely charged ions.
Ionic equation An equation in which only those ions and molecules taking part in the reaction are shown.
Ionic product of water The equilibrium constant for the ionisation of water, K_w.
Ionisation energy The first ionisation energy is the energy needed to remove 1 mole of electrons from 1 mole of gaseous atoms (under standard conditions).
Ion polarisation The distortion of the electron cloud of an anion by a (small and/or highly charged) cation.
Isotopes Atoms of the same element having different numbers of protons.

K

Kinetic theory Particles in liquids and gases are in constant movement and those in solids vibrate.

L

Lattice A regular repeating arrangement of atoms, ions or molecules in three dimensions.
Lattice energy The enthalpy change when 1 mole of an ionic compound is formed from its gaseous ions (under standard conditions)
Le Chatelier's principle When the conditions affecting the position of equilibrium are changed, the position of equilibrium shifts to minimise the change.
Ligand An ion or molecule with one or more lone pair of electrons which can form co-ordinate bonds with a central transition element ion.
Ligand exchange The replacement of one ligand by another in a transition element complex.
Lone pair A pair of electrons in the outer shell of an atom which is not bonded.

M

Mass number (*A*) The total number of neutrons + protons in a nucleus.
Metallic bond A bond formed in a metal by the attraction of metal ions to the sea of delocalised electrons.
Metallic radius Half the distance between the nuclei of two atoms in a metal.
Metalloid A substance whose electrical conductivity increases with an increase in temperature.
Molar mass The mass of a mole of a substance in grams.
Mole The amount of substance having the same number of specified particles as there are atoms in exactly 12 g of the carbon-12 isotope.
Molecular formula Shows the number of each type of atom in a molecule.
Molecular orbital An orbital formed by combining atomic orbital's from different atoms.
Mole fraction In a mixture of molecules (or atoms), the number of moles of a particular molecule divided by the total number of moles of all the molecules.
Monodentate Ligands which form one co-ordinate bond with the central transition element ion.

N

Non-polar A molecule with no separation of charge.
Nucleon number See Mass number.

O

Order (with respect to a particular substance) In a rate equation this is the power to which the concentration is raised. The overall order of reaction is the sum of these powers.
Oxidation Loss of electrons or increase in oxidation state.
Oxidation state/number A number given to a particular atom which describes how oxidised or reduced it is.
Oxidising agent A reactant that removes electrons (or increases the oxidation number) of another reactant.

P

Paramagnetic A substance which, when placed in a magnetic field, lines itself up with the field but does not retain its magnetism when the magnetic field is removed.
Partial pressure, *p* The pressure of an individual gas in a mixture of gases.
Permanent dipole–dipole forces Intermolecular force between molecules which have permanent dipoles.
pH $-\log_{10}[H^+]$

pH range (of indicator) The range of about 2 pH units over which particular indicators change colour.
pi bond A covalent bond involving sideways overlap of atomic orbitals.
pK_a $-\log_{10}K_a$
pK_w $-\log_{10}K_w$
Polar A bond or molecule in which there is a separation of charge.
Principal quantum level A region of space outside the nucleus of an atom which may contain a certain number of electrons up to a maximum number.

Q

Quantum Energy of fixed values only, which can be absorbed or emitted by an atom.

R

Radioactive decay The breaking up of an atomic nucleus, emitting particles in the process.
Radioactive isotopes Isotope of elements which have nuclei that break down spontaneously
Rate constant The proportionality constant in the rate equation.
Rate determining step The slowest step in a reaction mechanism.
Rate equation An equation showing the relationship between the concentrations of reactants which affect the rate of reaction and the rate constant.
Rate of reaction The amount or concentration change of a specific product or reactant per second.
Reaction mechanism A series of steps showing the stages in a reaction.
Real gas A gas which does not obey the gas laws especially at high pressure and low temperature.
Redox reaction A reaction where reduction and oxidation occur at the same time.
Reducing agent A reactant that donates electrons (or decreases the oxidation number) of another reactant.
Reduction Gain of electrons or decrease in oxidation state.
Relative atomic mass (A_r) The weighted average mass of the atoms of an element measured on a scale on which an atom of the carbon-12 isotope has a mass of exactly 12 units.
Relative isotopic abundance The proportion or percentage of a particular isotope in a mixture of isotopes.
Relative isotopic mass The mass of a particular isotope of an element measured on a scale on which an atom of the carbon-12 isotope has a mass of exactly 12 units.
Relative molecular mass (M_r) The mass of a molecule measured on a scale on which an atom of the carbon-12 isotope has a mass of exactly 12 units.
Resonance hybrid The actual structure of a compound is somewhere in between two or more forms.
Reversible reaction A reaction in which the products can be changed back to the reactants.

S

Salt bridge An inert support soaked in KNO_3 used to make electrical contact between two half cells.
Semi-conductor Substances which have low electrical conductivity at room temperature but increase in conductivity as temperature increases.
Shielding (screening) The ability of inner shells of electrons to reduce the effective nuclear charge on other electrons, especially those in the outer shell.
Sigma bonds Bonds formed by the end-on overlap of atomic orbitals.
Solubility product (K_{sp}) An equilibrium expression showing the concentration of each ion in a sparingly soluble salt solution raised to the power of their relative concentrations.
Spectator ions Ions which play no part in a reaction.
Stability constant The equilibrium expression showing the concentration of reagents involved in ligand exchange.
Standard cell potential The difference in the standard electrode potential of two half cells.
Standard conditions Pressure of 10^5 Pa and temperature 298 K.
Standard electrode potential The electrode potential of a half cell compared with a standard hydrogen electrode.
Standard enthalpy change An enthalpy change which takes place at pressure of 10^5 Pa and temperature 298 K.
Standard enthalpy change of atomisation The enthalpy change when 1 mole of gaseous atoms is formed from its element (under standard conditions).
Standard enthalpy change of combustion The enthalpy change when 1 mole of a substance is completely combusted (under standard conditions).
Standard enthalpy change of formation The enthalpy change when 1 mole of a substance is formed from its elements (under standard conditions).
Standard enthalpy change of hydration The enthalpy change when 1 mole of a specified gaseous ion is converted to an aqueous solution of ions (under standard conditions).
Standard enthalpy change of neutralisation The enthalpy change when 1 mole of water is formed from the addition of an acid to an alkali (under standard conditions).
Standard enthalpy change of reaction The enthalpy change when the specified molar amounts in the stoichiometric equation react (under standard conditions).
Standard enthalpy change of solution The enthalpy change when 1 mole of a substance is dissolved in excess water (under standard conditions).
Standard hydrogen electrode A half cell under standard conditions which has a Pt electrode dipping into 1 mol dm^{-3} H^+ ions in equilibrium with H_2 gas.
Stoichiometry The mole ratio of reactant and products shown in a balanced equation.
Strong acid/base Acids or bases which are completely ionised in solution.
Sublimation The direct formation of a solid from a gas or vice versa without a liquid state being formed.
Sub-shell Regions within the principal quantum shell where electrons have more or less energy.
Surroundings In a chemical reaction these include the container, the solvent and the air surrounding the reaction mixture as well as instruments dipping into the reaction mixture.
System The reactants and products in a chemical reaction.

T

Transition elements A d block element which forms stable ions with an incomplete d sub-shell.

V

Valence shell electron pair repulsion (VSEPR) theory The order of decreasing repulsion is lone pair-lone pair > lone-pair bond-pair > bond-pair bond-pair.
van der Waals forces Weak forces of attraction between molecules caused by temporary dipole-induced dipoles.

W

Weak acid/base Acids or bases which are partially ionised in solution.

Index

Key terms are in **bold** and are also listed in the glossary.